OVER OUR HEADS IN WONDER!

Science and the Sky

by Esther A. Yant

A Sourcebook of Bible Studies
with Science Activities

Cover Design by Christy Shaffer

ACKNOWLEDGMENTS

*Dedicated to my mother, Sara Andrews, and my husband, Howard,
without whose help this book would never have been written.*

Over Our Heads in Wonder
Science and the Sky

Copyright 1990, 2001, Esther A. Yant

Library of Congress Catalog Card Number: 2001087443
ISBN: 1-892427-02-8

Scripture translations are quoted by permission.
Verses marked (KJV) are from the authorized King James Version.
Verses marked (JB) are from the Jerusalem Bible. Copyright 1966. Darton, Longman,
& Todd Ltd. and Doubleday & Company, Inc.
Verses marked (TLB) are from The Living Bible. Copyright 1971.
Tyndale House Publishers, Wheaton, Illinois 60187. All Rights Reserved.
Verses marked (NKJV) are from the New King James Version.
Copyright 1979, 1980, 1982, Thomas Nelson Inc., Publishers.

Illustrations by Mark Kissner, Howard Yant, and Sara Sipe.

All rights reserved. No part of this publication may be reproduced or transmitted
in any form or by any means, electronic or mechanical including photocopy,
without the permission in writing from the publisher,
except for pages 75 - 84 which may be reproduced for teaching purposes.
Making copies of this book, or any portion for any purpose other than stipulated,
is a violation of United States copyright laws.

Published in the United States of America by:
BRIGHT IDEAS PRESS
877.492.8081

Second Edition

TABLE OF CONTENTS

INTRODUCTION 1

LETTER TO TEACHERS AND PARENTS .... 2 - 4

PART I — READINGS AND DISCUSSIONS

1. **WONDERS OF THE SKY** 5
   Sky Watching 6, 7
   Sky Exercises 9, 10
   Wonders of the Air 11, 12
   Wonder of the Wind 13, 14
   Catching the Wind 15
   Wonders of the Rainbows 16, 17

2. **WONDERS OF STORMS** 18
   Changing Clouds 20, 21
   Storm Story 22, 23
   Storm Watch 24
   A Storm, A Boat, and Jesus 25
   All in a Raindrop 26, 27
   All in a Snowflake 28, 29
   After the Storm 30

3. **WONDERS OF STARS** 31
   Nighttime Wonders 33
   Thinking About Stars and Planets 34
   Counting Stars 35, 36
   The Big Yellow Star 37
   The Sun and the SON 38
   Light of the World 39
   The Star of Bethlehem 40

4. **WONDERS OF PLANET THREE** 41
   From an Earthship Window I 43, 44
   From an Earthship Window II 45
   From a Spaceship Window 46, 47, 48
   View From Many Windows 49
   Last View of the Earth's Sky 50

TABLE OF CONTENTS

PART 2 – SKY ACTIVITIES 51

1. **ACTIVITY KITS** 52

2. **AIR EXPERIMENTS** 53
   Balloon Rocket, Candle Experiments, Air and Water

3. **BIBLE ACTIVITIES** 54, 55
   Rainbow Bible Bookmarks, Bible Stories and Verses

4. **CLOUDY DAY ACTIVITIES** 56
   Cloud Shapes, Recipe for Clouds
   Cloud Watching, Cloud Mobile

5. **EXPERIMENTS WITH LENSES** 57
   Hand Lens, Other Lenses, Binoculars.

6. **RAINBOW ACTIVITIES** 58
   Indoor Rainbows, Outdoor Rainbow
   Rainbow Poems and Pictures

7. **RAINY DAY ACTIVITIES** 59
   Raindrop Picture, Waterdrop Fun
   Waterdrop Game, Recipe for Rain

8. **SKY BOXES** 60
   Sky Activity Box, Sky Treasure Box

9. **SKY PICTURES** 61

10. **SKY WRITING** .... 62 – 65
    Sky Notebook, Sky Writing and Discussion,
    Sky Word Box, Sky Box of Science Words,
    Sky Diamonds, Sky Prayers

11. **SNOWY DAY ACTIVITIES** 66
    Snow Shapes, Snow Games, Snowflakes

TABLE OF CONTENTS

12. **STARRY NIGHT ACTIVITIES** .... 67 – 69
    Family Fun Under the Stars, Star Show Recipe
    Star Pictures, Pie in the Sky
    Star Verses in the Bible
    Special Star Activities for Schools

13. **STORMY DAY ACTIVITIES** 70
    Storm Experience, Storm Stories in the Bible

14. **SUNNY DAY ACTIVITIES** 71
    Sun Tricks, Sun Tag, Shadow Tag
    Sun Picture, Sun and Shadows

15. **WEATHER ACTIVITIES** .... 72 – 75
    Quickie Weather Report
    Newspaper/TV Weather Report
    Weather Chart (reproducible)
    Wind Speed Chart
    Thermometer Activities
    Measuring Rainfall
    Measuring Snowfall
    Kinds of Clouds
    Weather Chart

16. **WINDY DAY ACTIVITIES** 76
    Flying Discs, Parachutes

17. **REPRODUCIBLE PAGES** .... 77 – 84
    Cup of Wonder, Sky Box, Rainbow Picture,
    Snowflake Diamond, Sky Shapes,
    Balloon Picture, Through a Spaceship Window,
    Sky Designs.

INTRODUCTION

Abraham Lincoln said, "I can see how it might be possible for a man to look down upon the earth and be an atheist, but I cannot conceive how he could look up into the heavens and say there is no God."

Have you looked at the sky today? If you take time to look - really look - the sky comes alive with beauty and color. Suppose that the spectacular star show that unfolds nightly in the heavens, only occurred on one night every fifty years. Then on that night almost everyone in the world would be outside to see the sight of millions of stars blazing in the darkness. Yet because we can see that star show almost every night, most of us take it for granted.

How quickly children can sense the wonder and beauty of the sky and how easily this wonder can be directed into a joyful thanksgiving for God's creation, blessings, and great care for us. This book begins by showing God's greatness, power, and love in the vastness and beauty of the sky, the power of the weather, and the lights of the firmament. It ends by focusing on the Light of the world, Jesus, who came to Earth - the third planet in the solar system.

Why study the sky?

- The sky can always be observed - day or night, rain or shine.
- It is available everywhere - in cities or in the country.
- It's interesting to people of all ages and activities such as star watching can be shared by several generations.
- The sky can be experienced in many dimensions. You can see a rainbow in a single raindrop, a small hose sprinkler, a large waterfall, or the great expanse of sky.
- It can be experienced many different ways through color, smell, temperature, air pressure, and precipitation. It's always changing!
- The sky is free - you can study it with just your five senses. A few inexpensive materials are helpful but not necessary.
- Skywatching is a stimulus to creative expression. Inspiration from the sky encourages creative writing, drawing, and thinking.
- Studying the sky can open windows of reality to children. The sky can be a tool for learning about science, religion, and other subjects.
- Sky awareness can free children's minds from a TV mentality by getting them excited about the wonders of nature.
- Most importantly, the sky can teach us about God and His creation.

LETTER TO PARENTS AND TEACHERS

HOW DO I USE THIS BOOK?

Reading
The readings and activities are appropriate for children from kindergarten through fifth grade. The last section, beginning on page 41, is the most advanced. The beauty of this book is its adaptability for use with a variety of ages.

For younger students, read aloud to them or let them take turns reading small segments. The Bible passages, in particular, may be hard for them. Simplify the writing assignments as necessary. For older students, use more experiments and discussion questions. Let them read everything by themselves, either aloud to the family or to themselves.

There are twenty-five readings and more than fifty activities, which can be used in many different ways. A student can use this book as an individual science/Bible workbook, adding their own drawings and notes directly on the pages. Or the teacher can use it as a resource guide to developing her own unit on science/weather/Bible.

Student Notebooks
The Student Notebook Approach works terrifically with this book! Each student can keep their own notebook or journal throughout the course of this study. Use it to record information, write down questions, vocabulary words learned, experiments done, projects, artwork, Bible verses, and anything else relevant to the unit. Small children may prefer small journals! Choosing a little notebook that is cheerful and not overwhelming in size will give them the confidence and motivation to begin keeping records. This is a great lifelong habit to develop. (For more information on this method of education, research the Principle Approach and the Charlotte Mason "Living Books" approach.)

Unit Studies
Here are some great ideas to get you started:

Science - The science questions are good springboards to discussion. Children can look up sky topics they're interested in and write short reports or make displays to show what they've learned. Science vocabulary is sprinkled throughout.

Bible - Colorful rainbow bookmarks are a new way to find and remember stories and verses in the Bible. Bible storm stories can be read on stormy days and star verses on starry nights. *(See Bible Activities.)*

Math Skills - Measure amounts of snow and rain, conventional temperatures from Fahrenheit to Centigrade, computing the average weekly temperatures, or the total rainfall, etc. *(See Weather Activities.)*

LETTER TO PARENTS AND TEACHERS

Art Projects - Encourage awareness by observing the changing patterns in the sky. Take photos and draw pictures to record observations. Almost every page has room for student drawings.

Outdoor Activities - Whatever the weather, there are many activities for outdoor fun. *(See Windy Day and Sunny Day Activities for some examples.)*

Activity Kits - Set up learning centers for science experiments and art projects.

Weather Station - Make and use to record daily observations on a chart. *(See Weather Activities.)*

Sky Watching - Use during a break from studies: students run to the windows, observe the sky for two minutes, then return to their work. Use this to develop attention to details.

Research - Newspapers, encyclopedias, weather and sky atlases, TV weather reports, and the Internet are just a few of the many resources available for extra research. The sky's the limit!

Writing - Children can observe the sky each day and write in their sky journals. Poems and prayers can be written inside different sky shapes. *(See Sky Writing, Sky Diamonds, and Sky Prayers for many other ideas.)*

Handwriting - Scripture verses are good for handwriting practice. The handwritten verses can also be illustrated with rainbows, stars, etc.

Sample Lesson
1. Readings
2. Questions & Discussions
3. Bible Verses
4. Activity
5. Prayer

Using "All in a Snowflake" on pages 28 & 29 your lesson plan might look like this:

1. Read "All in a Snowflake."
2. Have a discussion about material read.
3. Make cutout snowflakes and mount them on colored paper.
4. On the back they can make a sky diamond, write a poem about snowflakes, or copy their Bible verse.
5. End in prayer based on what was learned in this lesson.

■ LETTER TO PARENTS AND TEACHERS

Daily Devotions
Many parents would like to have daily devotions with their children but they don't always know how to begin or what to use. The children may be of a wide age-span or may be bored with the materials currently being used. This book gives ideas for teaching about God anytime and anywhere.

Read this in a cozy corner on a stormy night, on a long car trip (while skywatching, of course!), or on a blanket out in the backyard. Inside or outside, winter or summer, this book offers a plethora of ideas for daily devotions to our Lord God!

- Parents can look up activities which fit the season and the weather. The Table of Contents is useful when you are looking for great stuff during the following times:
- Boring, rainy days when there seems like nothing to do.
- Sunny, summer days when you need new ideas for family fun.
- Travel times when there are new opportunities to explore nature and sky watch.
- Busy days when you only have just a few minutes for science or Bible.

"Minute Vacations"
Most of us have little snippets of time here and there. If you have two, or five, or ten minutes, you can take a "Minute Vacation" by concentrating on the sky. Use as many of the five senses as possible. Either focus on one small detail of the sky or the entire big sky picture, and ask yourself questions about it. What does a waterdrop reflect? Why can you still see the sun's light after it has disappeared over the skyline? When time permits, look up answers to your questions and record them in your sky journal or student notebook.

Sky watching can be enjoyed by children and adults of all ages and can be done at the playground, the kitchen window, during a car ride or anywhere you can see the sky. This quick vacation can be the pause that refreshes as you take a moment out of a busy day to see the sky and thank its Creator.

Blessings as You Discover the Sky!
However you choose to use Over our Heads in Wonder, we hope that you and your children will grow in your understanding of God's word and His incredible creation. Enjoy your study of God's marvelous work - the sky!

1. WONDERS OF THE SKY

Take a cup and fill it up...fill it up with wonder.

I wonder WHERE the wind goes in the sky,
WHEN the clouds are passing by,
WHY the sky looks black in space,
HOW the sun runs its race,
HOW raindrops grow...WHO makes the snow,
WHERE darkness goes...WHEN sunlight glows.
Only God knows all the WHYS of the mysteries in the skies.

In the great spaces of the universe,
there is a small blue-green planet,
spinning fast, flying past the stars.
I stand on this small planet,
on the edge of the sky,
watching the clouds float by,
watching the moon and stars,
from the windows of my earthship.

Of all the wonders in the universe,
Isn't life a great wonder?
I was born on this blue-green planet,
I can talk, hear, taste, smell, feel, and think.
I like to think big thoughts.
Why did God put me on this planet
with plants and animals all around me?
Are there people on other planets,
And weird-looking plants and animals?
What is your wonder? A wonder can be shared.

■ READINGS & DISCUSSIONS / 1. WONDERS OF THE SKY

Table of Contents

READINGS AND DISCUSSIONS
- Wonders of the Sky 5
- Sky Watching 7, 8
- Sky Exercises 9, 10
- Wonders of the Air 11, 12
- Wonders of the Wind 13, 14
- Catching the Wind 15
- Wonders of Rainbows 16, 17

ACTIVITIES

Air Experiments 53
BALLOON ROCKET
CANDLE EXPERIMENTS
AIR AND WATER

Bible Activities 54, 55
RAINBOW BIBLE BOOKMARKS
BIBLE STORIES AND VERSES

Rainbow Activities 58, 79
INDOOR RAINBOWS, OUTDOOR RAINBOW
RAINBOW POEMS AND PICTURES
RAINBOW PICTURE (reproducible)

Sky Boxes 60, 78
SKY ACTIVITY BOX, SKY TREASURE BOX
SKY BOX (reproducible)

Sky Pictures 61

Sky Writing 62, 63, 64, 65, 77, 81
SKY NOTEBOOK, SKY WRITING AND DISCUSSION
SKY DIAMONDS, SKY PRAYERS
SKY WORD BOX, SKY BOX OF SCIENCE WORDS
CUP OF WONDER (reproducible)
SKY SHAPES (reproducible)

Windy Day Activities 76
FLYING DISCS, PARACHUTES

Sky Watching

Open your eyes and look at the skies.
The sky is filled with many things-
raindrops and rainbows,
clouds and sunshine,
birds and airplanes,
stars and planets.
But most of the sky is empty;
it doesn't even have air in it.

We see the sky as blue,
filled with an ocean of air,
with big waves of wind that come rushing by.
But if we go up high in the sky in a spaceship,
the sky looks black and empty,
with only a few twinkling lights in it.

We can see only a small slice of the sky,
because the mountains, buildings and trees are in the way.
Only God can see the whole wide sky,
with all the different planets, meteors and asteroids.

When we see the big sky,
we have big questions about God and His universe.
We can ask a million questions about the sky,
And when we find the answer to one question,
God gives us more questions to ask.
Only God knows all the questions and answers.

Be a sky-watcher.
The longer you look, the more you see.
Watch the sky all day, and sometimes at night,
through windows, in your yard,
on hilltops, from a car, everywhere.

When you watch the sky, look for these things:
  ★ **Sky colors**.....*shades of blue, red, orange, yellow.*
  ★ **Flying things**...*birds, airplanes, balloons, kites.*
  ★ **Cloud shapes**...*animals, people, buildings, birds.*
  ★ **Skies spilling over**...*rain, hail, snow, sleet.*

■ READINGS & DISCUSSIONS / 1. WONDERS OF THE SKY

★ **Sun's activities**...*hiding behind clouds, warming the air.*
★ **Night sky wonders**....*moon, stars, planets, constellations.*
★ **Kinds of clouds**....*cirrus , stratus and cumulus clouds.*
★ **Rainbows**....*pieces of rainbows, full rainbows, double rainbows.*

What does the sky look like right now?
Close your eyes and tell about it in detail.
If you can't remember, go out and look at it with your eyes really open.
Isn't God's sky full of wonder and beauty!

> **RECIPE for a SKY**
> God made a big place..... With lots of space.
> He put in some air............ But not everywhere.
> A black bowl of stars...... A blue bowl of clouds.
> One big moon.......Hiding at noon.
> And sunlight.......... Hiding at night.

Questions to Think About:

Why is the sky blue?
Sunlight is a mixture of many different colors. As sunlight comes through the earth's air, the light rays break up and spread out. The short rays spread out more than the long rays. The blue rays are very short so they spread out more than the other colors. Most of the sky looks blue on a clear day because these blue rays are reflected back to us from all parts of the sky.

Does the sky change color if you climb up a mountain?
Whether you are looking from a mountain, or a ship, or a plane, the sky looks blue. However, the sky has many shades of blue. The blue of the sky gets darker as you climb a mountain, or go up in an airplane.

Why is the sky black when you get very far up?
It is black because there are no air molecules to scatter the blue color.
For example, astronauts see blackness in outer space where there is no air.

Here are some questions from the Bible. Can you answer them?
**"Which is the way to the home of the light, and where does darkness live?
Have you ever visited the place where snow is kept?
Has the rain a father? Job 38: 19,22,28. (JB)**

Do you know everything about light and darkness or snow and rain?
Who knows everything and can answer all questions about all things?

Sky Exercises

What can you do with a minute.... a minute with nothing in it?
Run to the window and look at the skies.
And take a minute to exercise.

> **Stretch your arms, and reach up high**
> **Pretend that you are touching sky.**
> **With your arms, swinging wide,**
> **Paint a rainbow from side to side.**

"Air-robics" – ★ First, open the window and take a deep breath of air. ★ Reach outside to catch a sunbeam, snowflake or raindrop, if you can. ★ Now, stretch up high, pretending to touch a star in the sky. ★ Then reach down and touch the earth or floor. ★ Pretend that you are holding a big sky brush. With your arms swinging wide, paint an imaginary rainbow. ★ Paint a big cloud in a funny shape. ★ Thank God for muscles which stretch and grow.

> **Now do the stretching with your eye.**
> **Looking low, looking high.**
> **Look at the tree roots growing downward,**
> **Look at tree shoots growing upward,**
> **Look up to the very top,**
> **Where the branches finally stop.**

One Minute Skyward Scan – ★ Look at a big tree, starting at the roots, and slowly looking all the way up the trunk, up the branches and to the top of the tree where it meets the sky. ★ Or let your eyes look up, up, upward to the top of a building or mountain that reaches up into the sky. ★ Thank God for the big sky, which unlocks our eyes from closeup things and stretches our eyeball muscles to look at faraway things.

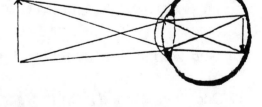

> **Take a picture of the sky,**
> **Pretend the camera is your eye,**
> **Take a picture in your brain,**
> **Of the clouds or sun or rain.**

One Minute Sky Photo – ★ Pretend you are taking a sky picture, and you yourself are the camera. ★ Place yourself in a good position to focus on the sky. ★ Then "snap" the picture, by closing your eyes and printing the picture on your memory. ★ Open your eyes and describe the sky picture to someone. ★ Thank God for eyes, which are better than a camera for catching pictures.

> Now stretch your thinking wide and high,
> Thinking of the earth and sky.
> Thinking big, thinking small,
> Thanking God Who made it all.

> Think of tiny things on earth,
> Which in the sky were given birth,
> Bits of stardust, bits of snow,
> Bits of rainbows, all aglow.

Mini Sky Pictures – ★ Take a pocket mirror outside, and hold it so it reflects a little piece of sky. ★ Find sky reflections in puddles and windows. ★ Look at a raindrop shining in the sunlight after a rainstorm. Does it reflect the sun or the sky? Sometimes you can find a mini rainbow in a raindrop. ★ Look at a soap bubble, which reflects the colors of the sun. ★ Thank God for tiny sky pictures all around us.

> Now think of big things out in space,
> The sun that makes its daily race,
> The moon that often shows its face.
> And planets orbiting in place.

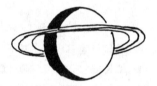

Five Minute Starview – ★ Go outside with your flashlight on a starry night, and look for constellations, shooting stars, moving lights of airplanes, and other unusual things. ★ Listen for night sounds and see if you can identify them. ★ Feel the night air. ★ Do this every night for a week, if you can. ★ Thank God for a skyful of stars, and a world full of wonders.

> Now you have stretched up very high,
> With your body, mind, and eye.
> You can never touch the sky,
> Even if you really try.

But God wants us to reach and grow...To stretch our thinking as we go.

"O God, here I stand....On this earth, on this land,
Teach me what there is to know...Help me so I stretch and grow,
Open my eyes so I can see...That you are always here with me. Amen."

Wonders of Air

An ocean of air surrounds our planet.
It is called the atmosphere.
This ocean of air has no color, smell, or taste.
We cannot see air around us,
but we can feel it as it blows against our faces.
Without air, there would be
no living plants and animals on this planet.

Air is a wonderful thing.
Did you ever wonder about how sounds
can come from many miles away?
Air carries sound, and without air,
this world would be a silent place.

Did you ever wonder about how the air
can hold raindrops and snowflakes?
At first, raindrops and snowflakes are tiny,
and so the air can hold them up.
But when they get bigger and heavier,
gravity pulls them down to earth.

Did you ever wonder about how the air
can hold an airplane up in the sky?
Air is much lighter than water.
But it can support birds designed by God,
and airplanes designed by man.
Scientists design airplane wings
to give lift to the plane,
and motors to give the plane power.

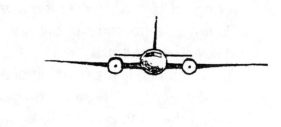

Air is one of God's wonders.
How can we learn about this world of wonders?
The whole universe is God's schoolroom.

We can learn by **studying**—science books and the Bible.
    We can learn by **looking**—sky watching and earth watching.
        We can learn by **thinking**—asking God to help us find the answers.

■ READINGS & DISCUSSIONS / 1. WONDERS OF THE SKY

As we study science and observe nature,
And as we read the Bible,
we are learning about the mysteries of sky and earth.
GOD'S WORLD AND GOD'S WORD,
both teach us about the wonders of the universe .

**"Meditate on God's wonders.
Can you tell how God controls them...?" Job 37:14, 15 (JB).**

Questions to Think About:

What is atmosphere?
The atmosphere is the ocean of air that surrounds our planet.
How deep is the atmosphere?
The atmosphere is hundreds of miles deep, but only the bottom few miles are thick enough to breathe.
What is air made of?
Air is made of several gases. The most important ones are nitrogen (78%) and oxygen (almost 21%). Water vapor and dust are mixed with the air in the sky.
How long can you live without air ?
A person can live for a month without food, and a week or less without water. However, a person can only live for a few minutes without air.
What is air resistance ?
Air resistance is the pushing of air against a moving object. When you ride your bicycle, you feel the air pushing against you. The faster you ride your bicycle, the harder the wind pushes against you.

Air Experiments (See Activities Section) – p. 53
Balloon Rocket, Air and Water, Candle Experiments

Wonders of the Wind

Up in the air so high . . .what can you see in the sky?
The sky is a room . . . the wind is a broom.
Every day up on high . . . the wind sweeps the sky.

Have you ever been up in an airplane,
and looked at a "sky room" full of clouds?
Have you ever thought about the wind
sweeping through the sky
like a giant broom?

Wind is moving air,
And it may blow gently so that
you can hardly feel it.
Or it can be a stronger breeze,
that blows smoke and leaves away.
Or it can blow so hard
that it whips up giant waves in the ocean,
and moves giant thunderstorms across the sky.

Where does wind come from?
Cold air is heavy, and hot air is light.
Cold air pushes the hot air up.
When the hot air goes up,
and the cold air pushes down, there is a wind.

**"And (God) brings the wind from his storehouse".
Jeremiah 10:13. (JB)**

God's storehouse of wind is the sky,
with cold and hot air going up and down, to and fro.

Wind is invisible, we cannot see it.
But we know it is there
because we can feel it blowing,
and see the clouds and trees
moving in the wind.
God's world is full of wonders and some of them, like the wind, are invisible,
so we can never see them.

READINGS & DISCUSSIONS / 1. WONDERS OF THE SKY

Questions to Think About:

Can you think of some invisible things in the sky?

1. Water vapor is the name for water in the air. We cannot see the water molecules in the air because they are too small. When millions of water molecules come together and form cloud droplets, then we can see them.
2. Dust particles in the air are usually too tiny to see. But the dust in the air helps to form the raindrops.
3. Sound waves, carried by the air, cannot be seen.
4. We cannot see radio waves, gravity, or electricity.

We cannot see God but we know that He is there. How do we know this? What are some other invisible gifts that God has given us?

Windy Day Experiments *(See Activities Section)*
Flying Discs, Parachutes – p. 76
Wind Speed Chart – p. 72

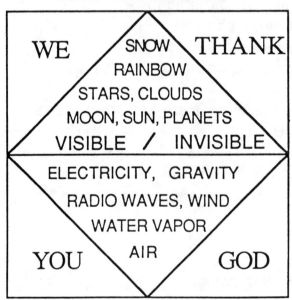

Catching the Wind

Once there was a boy.............who tried to catch the wind.
With a great big sack.............tossed over his back.
The North Wind was roaring........with a very loud sound.
The South Wind was exploring....looking all around.
The West Wind was soaring.........up in the clouds so high.
The East Wind was snoring...........asleep in the sky.
I'll catch the North Wind........the big boy said.
So he climbed up a mountain......where the sky was red.
And he opened up his sack.......when the North Wind blew.
He caught a piece of wind......and some snowflakes,too.
And the wind roared loudly.....inside the big boy's sack.
It wanted to get out.........It wanted to go back.
Back to the mountain......back to the plain.
Back to the place......where it was free again.

Can you catch the wind?
What if you had a big sack and filled it with the wind?
Would the wind roar loudly inside your sack?
No one is big enough to catch a big piece of the wind,
but we can catch small pieces of it.
We catch it in balloons and tires
We catch it with windmills and sailboats.
Can you think of other ways to catch the wind?

We cannot catch all of the mighty wind,
And even if we did, we wouldn't know what to do with it.
We cannot catch the wind or the stars or the sun.
All of these things are too big for us to imagine.
We cannot even imagine the power of the wind in the sky,
Or the sun power in the solar system.
How big and how far.....are the sun and each star.
How strong and how free.....are the wind and the sea.
How great and how high.....are God's stars in the sky.

God's ways are higher than our ways, and his thoughts higher than ours.
'For just as the heavens are higher than the earth, so are my ways higher than yours, and my thoughts higher than yours." Isaiah 55:9. (TLB)

Wonders of Rainbows

Where does God paint pictures for us to see?
The sky is like a beautiful blue mirror
which God paints with bright colors.
Each day at sunrise and sundown,
God paints a new picture with big brush strokes.
Sometimes the sky is fiery red and orange.
Sometimes there are soft yellows and pinks.

A rainbow is a beautiful sky picture
which God paints at special times.
Rainbow colors are soft, rain-washed colors,
not bright, sun-splashed colors.

What is the recipe for a rainbow?
   Mix together millions of raindrops,
   And millions of sunbeams,
   In a sky full of air.

As the sun shines through each raindrop,
the sunlight is broken up into different colors.
There are seven main colors,
and they are always in the same order:
red, orange, yellow, green, blue, indigo, and violet.
But these colors blend together
so that you only see about four colors clearly.

RAINBOW
GIFT FROM GOD
SUN'S BRIGHT COLORS
WRAPPED IN RAINBOWS

Look for rainbows any time
you see the sun shining through raindrops.
Rainbows come in all sizes.
You might see a tiny rainbow
in a raindrop, or a soap bubble.
Medium-sized rainbows can sometimes be seen
in a water sprinkler or water fountain.
Look for big rainbows in waterfalls,

READINGS & DISCUSSIONS / 1. WONDERS OF THE SKY

and the biggest rainbows of all – in the sky. God talks to us through His pictures in the sky. He paints colorful rainbows, and bright sunsets and sunrises to tell us of His love. Whenever you see a beautiful sky picture, think about God's love for you and everyone in the world.

* I John 4:8, 12, and 7 (JB)

Questions to Think About:

Why are some rainbows clear and some rainbows very hazy? If the raindrops are very small, they do not reflect the sun's colors very well, and the rainbow colors are hazy. If the raindrops are bigger, the colors are clearer.
Why are there two rainbows in the sky sometimes? This happens when the sun's colors are reflected two times inside the raindrop.
The first rainbow has the brightest colors and the red coloring is on the outside. The second rainbow has the lightest colors and the red coloring is on the inside
Can you find the end of the rainbow?
   Wouldn't it be fun to walk through a rainbow?
      Why do you think that God made rainbows?

"Can you spread out the gigantic mirror of the skies, as He does? Job 37:18 (TLB). Can we make sky pictures as God does? We can make jet trails in the sky with airplanes. But we can't make sunsets or sky-sized rainbows. Only God can.

Rainbow Activities *(see Activities Section)*
Indoor Rainbows, Outdoor Rainbows, Rainbow Poems and Pictures – p. 58
Rainbow Bible Bookmarks – p. 54

2. WONDERS OF STORMS

Take a cup and fill it up....
Fill it up with air,
Cold air, warm air, wet air, dry air.
Mix it all together....
You will get some stormy weather.
Falling raindrops, clouds spilling out,
Blowing snowflakes, wind rushing about.

What a surprise....meets your eyes....in the skies.
Wind is blowing....cloud are growing....it is snowing.
Rain is raining....water's draining....kids complaining.

On the next few pages, you will read about the wind and storms, rain and snow, clouds and other weather wonders.

Table of Contents

READINGS AND DISCUSSIONS
- Changing Clouds 20, 21
- Stormy Story 22, 23
- Storm Watch 24
- A Storm, A Boat, and Jesus 25
- All in a Raindrop 26, 27
- All in a Snowflake 28, 29
- After the Storm 30

ACTIVITIES

Bible Activities 54, 55
RAINBOW BIBLE BOOKMARKS
BIBLE STORIES AND VERSES

Cloudy Day Activities 56
CLOUD MOBILE, CLOUD RECIPE
CLOUD SHAPES, CLOUD WATCHING

Rainy Day Activities 59
RAINDROP PICTURE, RECIPE FOR RAIN
WATERDROP FUN, WATERDROP GAME

Snowy Day Activities 66, 80
SNOWFLAKES, SNOW GAMES, SNOW SHAPES,
SNOWFLAKE DIAMOND (reproducible)

Stormy Day Activities 70
STORM EXPERIENCE
STORM STORIES IN THE BIBLE

Weather Activities 72–75
KINDS OF CLOUDS
MEASURING SNOWFALL
NEWSPAPER/TV WEATHER REPORT
QUICKIE WEATHER REPORT
THERMOMETER ACTIVITIES
WEATHER CHART (reproducible)
WIND SPEED CHART

Changing Clouds

Wispy, misty, twisty clouds,
High in the sky, almost dry.
Playing games with the sun,
Always changing, on the run.

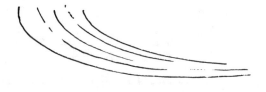

Lumpy, bumpy, humpy clouds,
Fluffy, puffy, scuffy ones.
Holding raindrops, maybe snow.
Always changing, on the go.

Gray, flat, heavy clouds,
Shutting out the sun and sky,
Dropping raindrops, maybe snow,
Always changing, on the go.

Clouds are changing, always changing.
Have you ever looked up at the sky
and watched cloud shapes
changing as the wind blows?
One minute a cloud looks like a crouching lion,
then a puff of wind turns it into a wooly bear.
Sometimes the wind pushes a cloud ship across the sky,
or makes a cloud giant marching in big boots across the sky.

What kind of clouds are up in the sky now?
Which clouds are rain clouds?
Even if we learn a lot about clouds,
we can't always tell what a cloud will bring.
Maybe it will blow away and bring sunshine,
Or maybe it will bring rain.
Even the weatherman doesn't know everything
about the weather. Only God knows.

"God is so great that we cannot begin to know him....He draws up water vapor and then distills it into rain, which the skies pour down. Can anyone really understand the spreading of the clouds and the thunders within? Job 36:26-29. (TLB)

Do you like stormy weather? Why does God make rainy days?

God gives us each day as a gift.
Each day is different and full of surprises.
Some days have rain and dark clouds.
Some days have sunshine and blue skies.

Are you happy with the weather
and the happenings of each day?
We can look at each day through dark glasses,
and see all the bad things in the day.
Or we can look at each day
through rainbow-colored glasses,
and see all the good things in the day.

How was your day today? Can you think of all the good things that happened today? God gave you a brand-new day, and He gave you eyes to see the beautiful world that He made. Think and thank God for today.

Questions to Think About:

Are there man-made clouds? You can see your breath as a cloud on a very cold day. And cars have little clouds coming out of the exhaust pipes on a cold morning. Jet trails are also clouds.
Can you go inside a cloud? When you walk through fog, you are walking through a low cloud. And sometimes when you fly in a plane, you go through clouds.
What does the inside of a cloud look like? It is foggy and misty inside a cloud because of the tiny water droplets in the air.
How are clouds formed? Clouds form when rising air expands and gets cool. Then the water vapor in the air turns into water droplets. These bits of water come together to form a cloud.

"Do you understand the balancing of the clouds with wonderful perfection and skill?" Job 37:16 (TLB) Even though we know a lot about clouds, we still have much to learn about their balance in the sky.

Cloud Activities (see Activities Section)
Recipe for Clouds, Cloud Mobile – p. 56
Cloud Shapes, Cloud Watching – p. 56
Kinds of Clouds – p. 74

Storm Story

(Mary, Bill, and Tom are playing in the kitchen while their mother cooks dinner. The TV and radio are making a lot of noise.)

TOM: Look, it's raining!
MARY: I see lightning in the sky!
(There is a loud noise. The TV, radio and lights go out.)
BILL: What happened?
MOTHER: The electricity went out because of the storm.
TOM: What can we do now?
MOTHER: Well, we can't do anything inside till the lights go on. Let's go and watch the storm from the porch.
(The four of them go out to the porch.)
BILL: I'm not scared; I like storms.
MARY: I got hit by a raindrop!
MOTHER: What can you see besides raindrops?
TOM: I see a bird flying backwards!
MOTHER: That's because the wind is so strong.
BILL: I see the trees bending backwards.
MOTHER: Listen to that strong wind. Where's it coming from?
MARY: Maybe it's coming from China or the North Pole or Hawaii.
TOM: The raindrops are dancing on the street. I want to go out and walk in the puddles with bare feet.
MOTHER: Wait till the lightning and thunder finish, then you can go out.
(Just then the TV and lights go on.)
BILL: The TV's on. Let's go watch it.
MARY: I'd like to stay out here, and then go for a walk in the rain.
TOM: Me, too.

Questions to Think About:

Why did the electricity go off during the storm?
That was the power of the lightning hitting the electrical wires somewhere. There are many kinds of power in the world. **What kind of power is used in radios and stoves?** Electric power.
What kind of power do you use when you think up a story or do math problems? Mind power.

There is battery power, in flashlights and toys.
There is muscle power. We aren't as strong as Superman, but we have good strong muscles.
There is imagination power. We can make a picture of something in our mind, and then draw it on paper.
Can you think of other kinds of power – in the earth, in the sky, anywhere?

What is the strongest kind of power in the world?
God's power is the strongest, because He made sun
and wind, lightning and electricity, and all things.
Without God's power, everything would stop:
electric machines, the wind, the sun, the people.....
everything in the world!

Superman is a strong guy . . . He can even fly.
Machines are powerful, too . . . That is really true.
My mind is very strong . . . My thoughts are very long.
But God has all the power . . . He can make a flower.
Or a mountain, or a tree . . . Or a person, like me.
"Power belongs to God." Psalm 62:IIb (NKJV)

There is power in your muscles,
There is power in your mind,
There is power in the wind,
There is power in the storms,
God's power is the most powerful of all!

"We cannot comprehend the greatness of His (God's) power. For he directs the snow, the rain and storm to fall upon the earth. Man's work stops at such a time, so that all men everywhere may recognize his (God's) power. The wild animals hide in the rock . . . From the south, comes the rain, from the north, the cold. God blows upon the rivers and even the widest torrents freeze. He loads the clouds with moisture, and they send forth his lightning . . . Listen, O Job, stop and consider the wonderful miracles of God." Job 37: 5b-7, 9-11,14. (TLB)

Storm Activities *(See Activities Section)*
Storm Experience, Storm Stories in the Bible – p. 70

Storm Watch

A storm is very exciting to watch.
Have you ever been up in an airplane during a storm,
flying through the storm clouds,
with the plane bouncing up and down?
Have you ever watched a storm from your porch or open garage?

The closer you are to the storm, the more exciting it is.
If you stand under a roof or sit on a porch, you can see
the raindrops dancing and the clouds changing minute-by-minute.
You can count the seconds between lightning strikes and thunder noises.

After a rainstorm, you can walk through puddles with old tennis shoes.
Look at the tiny silver raindrops on trees, plants, and buildings.
Watch the little streams of waters coming together.
Think about the little puddles and streams
making bigger streams and running into rivers and lakes.
Think about big rivers and lakes running into the ocean.

Questions to Think About:

What is a weather forecast? A weather forecast tells what the weather will be like. This helps people decide what to do for the day.
Who especially needs to know about the weather? There are some people who have to know about the weather so they can plan their work. Airplane pilots have to know what the weather will be like before their airplanes take off. Fishermen have to know whether there will be stormy weather at sea. Farmers use forecasts to plan their planting and harvesting.
What does a forecaster do? A forecaster collects facts from weather stations all around the world. Weather facts also come from satellites. Forecasters make charts to show the weather patterns. They use these charts and facts to make weather forecasts. These forecasts are on TV, radio, and in the newspaper.

Weather Activities (*See Activities Section*) pp. 72-75
Kinds of Clouds, Measuring Rainfall, Measuring Snowfall,
Newspaper/TV Weather Report, Quickie Weather Report,
Thermometer Activities, Weather Chart, Wind Speed Chart.

A Storm, a Boat, and Jesus

Mark 4:35-41
Note: *This story can be dramatized by assigning parts and using sound effects to make wind and wave noises. For example, someone could make wind noises by whistling and blowing. The wave noises can be made with water splashing in a jar. The spoken story and sound effects can be recorded on a tape recorder.*

Characters: First Narrator, Second Narrator, Jesus, Disciples.

First Narrator: One evening Jesus got into a boat with His disciples to go over to the other side of the lake. So, they started out in the boat, and as they were going across the lake, a storm came up.

Second Narrator: *(Make wind and wave noises).* The wind was blowing, and the clouds were growing! The waves were getting high! The boat was not so dry!

First Narrator: Jesus was asleep on a pillow in the stern of the boat, and the disciples woke Him up.

All Disciples: Master, Master! Do you not care? We are going down!

First Narrator: Then Jesus got up, and he spoke to the storm:

Jesus: "Peace, be still!"

Second Narrator: *(Storm sounds suddenly stop.)* The wind stopped blowing, and the waves stopped growing. The water became very calm. Out came the sun, the storm was done.

First Narrator: Jesus spoke to His disciples who had been so worried during the storm.

Jesus: Why were you so frightened? Have you no faith?

First Narrator: But they were surprised and afraid, and said:

All Disciples: Who is this? Even the wind and the waves obey Him.

To Think About:

The disciples used their muscle power to keep the boat going. They used their mind power to worry about the storm. But their boat was sinking because of the powerful storm. Only God's power could save them.

All In A Raindrop

What is the story of one tiny raindrop?
In the air are little bits of water,
so tiny that you can't see them.
As the air climbs up into the sky and gets cooler,
a cloud is formed.
Inside the cloud, the tiny water droplets push
and shove against each other.
Then they come together, and a raindrop is born.
When the raindrop gets big and heavy,
it falls to the ground.
It might fall on a tiny flower, a big tree, or a rock.

How many raindrops are in a storm?
It would take a long time to count all the raindrops in a storm,
and then we might run out of numbers.
Have you ever been to the ocean, or a big lake?
Think how many waterdrops it would take to fill an ocean or a lake.

Can you count the raindrops in the sky?
Can you count the birds that fly?
How many leaves are on a tree?
How many waves are in the sea?
How many grains of sand.
Can you hold in your hand?

How many raindrops fill up a puddle?
My head's in a muddle....
Thinking of millions and billions,
And trillions and zillions.
How many stars are in the sky?
Can I count them if I try? Not I!

What if God made only five flowers, ten rainbows or twelve raindrops in the whole world? Instead He made billions and trillions of flowers, rainbows, and raindrops. We can't count God's riches even if we use all the numbers in the world.

"You (God) visit the world and water it. You load it with riches."
Ps. 65:9a (JB)

Questions about Rain

How big are raindrops?
If the raindrops are very big, they are the size of small peas. But there are small raindrops, too. These are so small that you need a hundred to make an inch on your ruler.

What shapes do raindrops have?
Raindrops do not have a tear drop shape. When they are in a cloud, they are often round. But when they start to fall, raindrops look like hamburger buns, egg shapes, or other shapes.

What happens before a rainstorm?
There are many signs that rain is on the way. For example, bugs fly low and bite more often. Birds chirp more loudly, and cows huddle together. Some plants close up their flowers, and turn over their leaves. People with arthritis complain that their joints hurt more before a rainstorm.

Where does rainwater go after it rains?
Rainwater runs down drains and into underground pipes in cities. In the country, rainwater goes into rivers, streams, and lakes. And some of it goes into underground streams and wells. We catch this water by digging canals, making dams, digging wells, and many other ways.

Rainy Day Activities *(see Activities Section)*
Measuring Rainfall – p. 73
Raindrop Picture, Recipe for Rain – p. 59
Waterdrop Fun, Waterdrop Game – p. 59

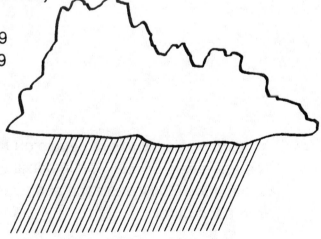

All in a Snowflake

Two snowflakes were falling through a frosty cloud.

"I am the most beautiful snowflake in all the world," one snowflake said to the other one.

"No, you aren't," argued the other one.

At that moment, the snowflakes fell through the bottom of the cloud and came out into the sunlight. All around them were other snowflakes shining in the sun.

"We're all beautiful," the two snowflakes agreed, as they continued falling and falling through the sky. Then they landed on the bright red mitten of a boy named John.

John looked at them carefully. "How funny!" he exclaimed. "This snowflake looks like a wheel, and that one is like a space ship."

John didn't hear the snowflakes talking. "I am not a wheel; I am a snowflake star," one snowflake said.

"You are a little snow picture, just like me," the other one said.

Have you ever caught snowflakes on your mitten and looked at them carefully?

All snowflakes have six points, but each one is different. Out of billions of snowflakes in a snowstorm, none are the same. Only God could make such tiny snow pictures so perfectly.

SNOWFLAKE
SIX-POINT STAR
GOD'S HANDIWORK
EACH ONE DIFFERENT
SPECIAL, BEAUTIFUL
SNOWSTAR

God made billions of people on this earth, but there is no one just like you. You are a very special and important person, and so is every other person that God made. God loves each and every person, even though we are all different from each other.

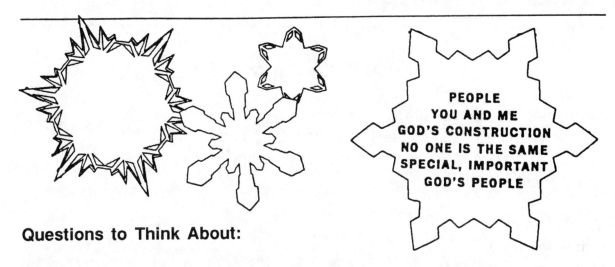

Questions to Think About:

Why did God make people so different, one from another?
Why are brothers and sister so different?
Why are friends so different, even though they like to do things together?
What would happen if everyone was the same?
Does God want us to love people that are very different from ourselves?

"Love your neighbor as yourself." Leviticus 19:18, Matthew 19:19. (TLB)
Your neighbor is anyone who lives on this planet. That's everyone!

Questions About Snow:

How do snowflakes form? Waterdrops in high clouds usually freeze into tiny bits of ice. As the bits of ice fall, more water freezes on them. These ice bits become ice crystals which come together to make snowflakes.
Why are some snowflakes like feathers, and some like hard sand? As snow falls, the flakes may stick together in big feathery clumps. Or the snowflakes may melt a little and then freeze again into hard bits. The snowflakes change as they fall through cold and warm air on their way to the ground.
Why does snow fall to the ground when it is so light? Snow is heavier than air, so it falls through the air to the earth.
Are all snowflakes shaped like stars? No. Often the water droplets freeze into snow crystals like flat stars, but sometimes they have other shapes. And the star-shaped snowflakes often break or change their shape as they fall to the earth.

Snowy Day Activities *(see Activities Section)*
Snow Games, Snow Shapes, Snowflakes – p. 66
Measuring Snowfall (Weather Activities) – p. 74

After the Storm

When the hard storms come,
with heavy thunder and lightning, we may be afraid.
When tree branches break, and floods are starting,
we may feel fearful.

But after these storms,
comes the wind,
breaking up the clouds,
comes the sun,
making a seven color rainbow.
After a storm, we are surprised by God's sunlight.

After a time of problems,
we are surprised by God's light and love.
Sunlight after a rainstorm,
Love after a quarrel,
Happiness after sadness.
God always gives us His light and His love.

The weather is changing, always changing,
Our feelings are changing, always changing,
God's love is everlasting, never changing.

When our eyes are opened, we will see that God is always with us,
In our problems, in our pain, in our joy, in our fun.

"I (God) have loved you, O my people, with an everlasting love."
Jeremiah 31:3. (TLB)

Questions to Think About:

Have you been in a big storm? What was it like? What do you do when you are afraid?
Talk over worries and fears with someone in your family. Then pray about each fear, asking God to fill your mind up with joyful thoughts instead of fear thoughts. There are many Bible verses which promise God's help in time of trouble and fear. Look up Psalm 46:1, Psalm 4:8, and Isaiah 43:5.

3. WONDERS OF STARS

Take a cup and fill it up....
Fill it up with sky stars,
Yellow stars, blazing white stars,
Supergiant stars, dwarf stars.

Or take a cup and fill it up....
Fill it up with earth stars.
Snow stars, flower stars,
Tree stars, sea stars.

How many, many stars there are!
And they are mostly very far!

■ READINGS & DISCUSSIONS / 3. WONDERS OF STARS

Table of Contents

READINGS AND DISCUSSIONS
- **Nighttime Wonders** 33
- **Thinking About Stars and Planets** 34
- **Counting Stars** 35, 36
- **The Big Yellow Star** 37
- **The Sun and the SON** 38
- **Light of the World** 39
- **The Star of Bethlehem** 40

ACTIVITIES
- **Bible Activities** 54, 55
  - RAINBOW BIBLE BOOKMARKS
  - BIBLE STORIES AND VERSES

- **Starry Night Activities** .... 67 – 69
  - FAMILY FUN UNDER THE STARS
  - PIE IN THE SKY
  - SPECIAL STAR ACTIVITIES FOR SCHOOLS
  - STAR PICTURES
  - STAR VERSES

- **Sunny Day Activities** 71, 84
  - SUN AND SHADOWS
  - SUN PICTURES
  - SUN TAG
  - SUN TRICKS
  - SKY DESIGNS (reproducible)

Nighttime Wonders

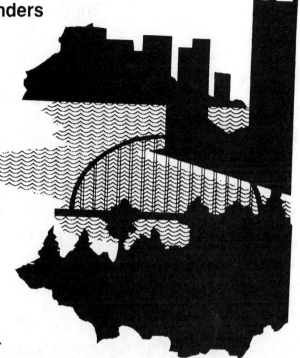

What is the best way to see the stars?
Take your flashlight and go outside
on a clear, moonless night. Find a place
where you can see the stars clearly.
If you live in the city,
a balcony or roof of a tall building
are good places for star watching.
You can also look for stars on a hillside,
in a park, or in your own back yard.
If you close your eyes for a few moments,
and then open them,
you will see the stars more clearly.
For the best results,
lie on your back on a blanket and look up.

As you look at the stars, forget about the earth and yourself. Just think about whatever is up in that great black sky. Think of some big questions. Use your flashlight as a pointer and look for these sky wonders :

Planets – The planets will look like extra-bright stars, but they are not. Jupiter, Saturn, Venus and Mars are the planets nearest the earth. Mars is one of the brightest objects in the sky.

Shooting stars – These are called meteors, and they look like a bright line of fire that flashes for a moment across the sky. Meteors are pieces of stars that have broken away and come into the earth's air where they catch fire.

Constellations – Look for groups of stars in a pattern. The Big Dipper is a constellation that is easy to see all year long.

Galaxy – Find a patch of light which looks like a milky river. This is the Milky Way, a galaxy which has billions of stars in it. There are millions of galaxies in the universe, and each one is an island of stars, dust, and gases.

There are many Bible verses about stars, such as Psalm 148:1, 3 and Psalm 8:1-4. You can read them by flashlight outside on a starry night. Or you can look them up when you are inside.

Thinking About Stars and Planets

How far away are the stars and planets?
Some of them look so near
that you could reach out and touch them.
But they are so far away
that we can't travel to them,
even in a rocket ship.

Would it be fun to live on a planet?
Do you think that God put other people on planets?
We don't know, but someday we may find out.
if we visit other stars and planets.

When you look up at the sky,
do you wonder if God is up there?
God is not in the sky; He is everywhere.
A long time ago, people thought
that God was up on a high mountain, or in the sky.

But the Bible says that God is with us,
wherever we are.
If you lived on a star, or under the ocean,
or on top of a mountain,
God would still be close to you.

When do you feel close to God?
Maybe you feel close to God
when you are outside, enjoying God's world.
Or maybe you think about God's love for you
when you are with your family or in church.

Remember that God is always close to you
whether you are scared and lonely, or happy and peaceful.
He will always hear you when you talk to Him.

The Bible says: **"Call to me, and I will answer you."** Jeremiah 33:3 (JB)

Counting Stars

How many stars are in the sky?
With gigantic telescopes,
scientists count about 100 billion stars
that can be seen.
And there are even more than that.
Who can count all the stars in the sky,
and who knows all their names?
Scientists are always counting stars,
naming stars and finding new ones.
But only God knows how many stars there really are.

**"He counts the number of stars;
He calls them by name." Psalm 147:4 (NKJV)**

How big is the sky with all the billions of stars in it?
A flashlight will help you learn
about the bigness and greatness of God's sky.
If you take a flashlight and point it at things on the ground.
The light always bounces back from each object.
If you point the flashlight up into the black sky,
you can't you see the light bouncing back. Why?
Instead of bouncing back, the light travels on and on
into the big distances of God's universe.
How far will it go? Who knows!
God's sky is bigger than we can imagine.

If you built a wall in the sky, it would never end because there would always be sky beyond it.
The sky is too large for us to measure. In this gigantic sky, God placed millions and billions of stars. God made billions of people who live on this planet Earth. And He loves and cares for each person in the world.

God's love for each one of us is bigger and deeper and wider than the sky. We can't measure God's love for us, just as we can't measure the sky. We can't count all the good things that God gives us, just as we can't count all the stars.

■ READINGS & DISCUSSIONS / 3. WONDERS OF STARS

Questions About Stars:

What are stars made of?
Stars are made of very hot gases. The main gas is hydrogen.

How are stars born?
Stars are born out of a huge cloud of dust and gas in space. When the gas and dust whirl together, a giant ball is formed. Pressure builds up inside the ball, and the temperature gets hotter. The inside of the ball starts to glow as the temperature rises. The hydrogen begins to turn to helium and produce great amounts of energy. This energy heats the outside of the ball, and it starts to shine. A star is born.

Why does a star die?
Stars use their own gases as fuel to produce light and heat. When they use up all their fuel, then they die. But stars are very big, and they can shine for millions or billions of years before they run out of fuel and die.

Why can't we see the stars all the time?
The stars send out light all the time, but we can't always see them. During the day the light from the sun is so bright that it hides the light from the stars. At night, the bright lights from cities often make the stars hard to see. That's why its better to do star-watching away from bright lights.

Why do the stars seem to be moving across the sky?
The earth is a planet in the solar system, so it moves through space with the sun. The earth spins around once a day, and also moves in an orbit around the sun once a year. Because the earth is always moving, the stars seem to be moving across the sky. The stars seem to rise and set at different places in the sky during different seasons of the year.

Star Activities *(See Activity Section)* p. 67-69
Family Fun Under the Stars
Star Show Activity
Star Pictures
Pie in the Sky
Star Verses in the Bible
Special Star Activities for Schools

The Big Yellow Star

What is the star that is closest to the earth?
This star is very big – a million times bigger than earth.

Did you guess the sun?
The sun is our own star
which warms our own group of planets.
It gives us just a tiny part
of the tremendous amount of heat
and light that it produces.
It is so bright that when it shines,
other stars in the sky just disappear.

The sun is an enormous ball of very hot gases.
These gases explode and burst out into space.
The temperature is about 10,000 F. degrees
on the outside of the sun,
And 27,000,000 degrees on the inside.
All living things store up power from the sun.
The sun gives light, heat, and life.
Aren't you glad that God made the sun with its powerful heat and light?

"And God made two huge lights, the sun and moon,
to shine down upon the earth." Genesis 1:16. TLB)

Questions to Think About:

What are sunspots? They are dark spots on the sun, but no one really knows what causes them. Maybe the gases break through the sun's surface, and cool, so they shine less brightly. Even so, the sunspots are not very dark. They just look dark in the brightness of the sun.
Will the sun get cold some day? Yes, but we don't have to worry about it. That won't happen for millions of years.
Is the sun one of the largest stars? Compared to other stars, the sun is only medium-sized. It is called a yellow dwarf. Other stars have a diameter 1,000 times as large as the sun.
Does the sun spin around, the way the earth spins? Yes, but it goes around much more slowly. The earth spins around once each day. The sun spins around once in about three weeks.

The Sun and The SON

The sun is God's special gift to our planet and the solar system.
The SON is God's special Gift to each of us on this planet.
Jesus Christ, Son of God, came to earth for you and me.
**"For God so loved the world,
that He gave His only begotten Son." John 3:16 (NKJV)**

The sun is the center of the solar system.
The SON is the center of our lives.
**"We are God's work of art,
created in Jesus Christ." Ephesians 2:10a (JB)**

The sun will become old and very cold.
The SON is everlasting, never changing.
**"Jesus Christ is the same yesterday,
today, and forever." Hebrews 13:8. (NKJV)**

The light of God's sun comes from 93,000,000 miles away.
The love of God's SON is right here with us now.
Sometimes we feel unhappy and faraway from His love.
But Jesus is always with us, no matter how we feel.
**"And know that I am with you always;
yes, to the end of time." Matthew 28:20b (JB)**

Questions to think about:

What are three things that you know about the sun?
What are three things that you know about God?
Why is the sun so important to us?
Why is God's Son so important to us?

Sunny Day Activities (see Activities Section)
Sun Tricks, Sun Tag, Shadow Tag, Sun Picture, Sun and Shadows – p. 71

WARNING:
Do not look directly at the sun!
The sun's rays are very powerful, and they can hurt your eyes.

The Light of the World

The sun gives us light and warmth.
It brings color in our world, and makes everything grow.
But too much sun can kill plants, animals, and people.
Venus is an example of a planet that has no life on it
because it is too close to the sun.

We cannot depend on the sun for life.
Only God can give us what we need.
God gives us life and love,
and we cannot live without Him.
Like a tree stretching up to the sun,
we can grow upward toward God.
God gives us His love and His light,
In Jesus Who is the LIGHT of the world.

Read Matthew 20:29-34. In this story, Jesus opened the eyes of two blind men. When they opened their eyes they were surprised by light.

★ Light from the sun, shining into their eyes, so they could see colors and shapes.

★ Light from Jesus, shining into their hearts, so they could know joy and love.

Jesus says, **"I am the Light of the world; anyone who follows me will not be walking in the dark: he will have the light of life." John 8:12.(JB)**

Questions to Think About:

What is light? No one really know what makes up light. We know that light travels in waves, and that it seems to be made of tiny particles. Scientists call those particles photons.

Where does light come from? Light comes from stars, such as the sun in our solar system.

How fast does light travel? Light travels very fast, about 186,000 miles a second. It takes about eight minutes for the sun's light to reach us.

Why is it dangerous to look at the sun? The sun's light is very powerful, and our eyes are easily hurt. You should never look directly at the sun, even with sunglasses. Your eyes could be hurt or blinded.

■ READINGS & DISCUSSIONS / 3. WONDERS OF STARS

The Star of Bethlehem

> High above our planet bright,
> A special star glowed in the night,
> Other stars were shining bright,
> Other planets were alight,
> But one star above the earth,
> Told about our Saviour's birth.

When Jesus was born on Christmas Day, God sent a very bright and beautiful star to tell people about his birthday.

Before Jesus was born, the world was a dark and lonely place. People didn't know much about God, and they didn't love and help each other very much. Jesus brought God's light and love to the world. He showed us how to love and help each other.

**"The people that walked in darkness have seen a great light."
Isaiah 9:2 (KJV)**

All the stars in the world remind us of that wonderful star that shone on Christmas night . . . **THE STAR OF BETHLEHEM.**

There are stars in the sky, the oceans, and on the earth.
    Of course, you have seen stars in the sky. But have you ever seen
        a star-shaped leaf,
            or a star-shaped flower,
                or a starfish in the ocean,
                    or a star-shaped snowflake?
God made many different star shapes in this beautiful world. (See p. 84)

Stars remind us that Jesus is the Light of the world.

4. WONDERS OF PLANET THREE

Take a cup and fill it up . . . fill it up with wonder.

I wonder why God made me . . . why He loves this Planet Three.
Why is this a special place . . . Planet Three in solar space.

Will I ever take a trip . . . on a starbound rocket ship?
Are we moving as we stand . . . on this earth, on this land?

If you look up in the sky . . . are the planets spinning by?
Are the stars flying past . . . is the moon rising fast?

Only God knows all the whys . . . of the mysteries in the skies.
Only God can tell us all . . . about this tiny blue-green ball.

Earth, the tiny blue-green ball . . . what's its secret all in all?
Do you know?

■ READINGS & DISCUSSIONS / 4. WONDERS OF PLANET THREE

Table of Contents

READINGS AND DISCUSSIONS
From an Earthship Window I 43, 44
From an Earthship Window II 45
From a Spaceship Window .... 46 – 48
View From Many Windows 49
Last View of the Earth's Sky 50

ACTIVITIES
Air Experiments 53
BALLOON ROCKET
CANDLE EXPERIMENTS
AIR AND WATER

Bible Activities 54, 55
RAINBOW BIBLE BOOKMARKS
BIBLE STORIES AND VERSES

Experiments With Lenses 57
HAND LENS, OTHER LENSES, BINOCULARS

Sky Pictures 61, 82
SKY PICTURES, BALLOON PICTURE (reproducible)

Sky Writing .... 62 – 65
SKY NOTEBOOK
SKY WRITING AND DISCUSSION
SKY DIAMONDS, SKY PRAYERS
SKY WORD BOX, SKY BOX OF SCIENCE WORDS
THROUGH A SPACESHIP WINDOW (reproducible)

From An Earthship Window – Part I

Right now, you are traveling through space,
watching the stars and planets go past.
The name of your spaceship is "Earthship".
Can you feel the earth moving under your feet?
What can you see from the windows of your "Earthship"?

We can see God's beauty in sky pictures.
Big sky pictures...sunsets and rainbows, and starry nights.
Little sky pictures.....snowflakes, and sparkling raindrops.
God talks to us through sky pictures,
But all the colors in the rainbow,
Can't describe all the colors of God's love.

We can see God's power in the world and sky.
Sun power spilling into flowers,
Wind power sweeping out the sky.
We can see God's greatness in the universe.
Stars shining through millions of miles in space.
Planets whirling around different suns.

We can learn about God from watching the sky.
But the sky doesn't teach us everything about God's love,
Because God's love is wider, deeper and higher than the sky.

God's love is for all around us,
Just as the air is all around us.
The air is invisible; we can't see it.
But we know it is there because
it fills up balloons and holds up kites.

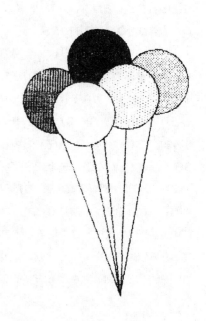

God is invisible, too. We can't see Him.
But we know God is there
because He shows us His love
in thousands and millions of ways.

Think of the air in a balloon.
Without air, a balloon is wrinkled and empty.
God's love is like the air, and when it fills us up,
then we can love other people.

**Let us love one another,
since love comes from God. I John 4:7 (JB)**

Think of a kite in the sky.
The kite is help up by the wind,
and without the wind it would fall down.
So, we, too, are held up by God's love,
even in time of trouble.
God is love. I John 4:8. (JB)

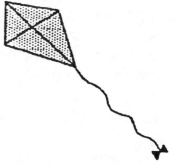

> **Can you see the wind as it goes by?
> Can you see the highest star in the sky?
> Can you see love? Can you see joy and fun,
> That God gives to you, and everyone?**
>
> **I cannot see the wind when it is passing by.
> I cannot see the stars, which are so very high.
> I cannot draw a picture, even if I try,
> Of love and God, and things invisible to my eye.**

Questions to think about:

There are some things we cannot see because they are hidden from us.
Can you think of something in the sky that is sometimes hidden behind clouds? Can you think of things that are hidden because of the sky's darkness or brightness?
There are some things we cannot see because they are too far away.
Can you think of some things in the sky that are too far away to be seen?
There are some things we cannot see because they are invisible.
Can you think of some things in the sky that are invisible?
There are some faraway things we cannot see clearly with our eyes, but we can see clearly with lenses, such as telescopes and binoculars.
Why do things look different when you look at them through a glass lens?"
Light travels at different speeds through different things, and it goes more slowly through water or glass than it does through air. When you look through a glass lens at something, the light slows down, and changes direction. This makes the object look either bigger or smaller.

From an Earthship Window – Part II

***Sky in sight . . . moon is bright,
Stars at night . . . God's delight.***

We can look out our earthship window and see a sky full of stars.
But there are many stars we cannot see
because they are too far away.
When we want to see something faraway,
we use a telescope or binoculars.
Have you ever looked at the moon
or the stars through a telescope or binoculars?
A telescope has glass lenses
which make stars and planets look bigger.

Our earthship has big windows to look out at the sky.
But there are also small windows
to look at small things on earth.
When we want to see something really small,
we use a hand lens or microscope.
Have you ever looked through a microscope
or through a hand lens?
Sand, sugar, and salt look like tiny jewels
when we see them with a hand lens.
Objects look bigger and clearer.

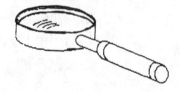

Telescopes, binoculars, microscopes,
and hand lenses help us to see more clearly.
But we cannot see very clearly,
even when we look through a glass lens.
And there are many things we cannot see.
Only God sees everything, and knows everything.
Someday God will show us things that we can't understand now.
Can you think of some questions that no one but God can answer?

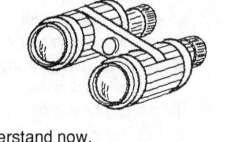

"For now we see through a glass darkly; but then face to face. Now I know in part, but then I shall know, even as I am known." I Cor. 13:12 (KJV)

Activities (See Activity Section)
Experiments with Lenses – p. 57, Sky Diamonds (Big/Small) – p. 64

From A Spaceship Window

Pretend you are traveling in a spaceship, orbiting the earth. Looking through your spaceship window, you can see the earth unfolding.

Down on the Earth below . . . what does it show?
The Earth has a floor . . . from shore to shore.
With a blue-green rug . . . that doesn't fit snug.
With wrinkles and crinkles . . . from mountains and hills.

As you orbit the Earth, even the biggest mountains look like small folds in the Earth's crust. The largest rivers are like blue threads, and the biggest lakes like tiny silver mirrors. You cannot see any cities, electric lights, or roads.

Now you are leaving Earth's orbit.
As you zoom into space,
you look at the black sky with millions of tiny lights.
Chunks of rock float past your window.
Everything is quiet,
except for the hum of the machines
in your spaceship.
The moon glows softly
and gets bigger and bigger.
As you look back,
you see the Earth growing
smaller and smaller.
First it is the size of a basketball,
then it is no bigger than a baseball.
And at last it is the size
of a tiny blue-green marble.
Holding your fingers up to the window,
you measure the Earth
between your thumb and forefinger.
It isn't very big is it?

When you travel through space.....away from Earth's place,
At first you can see...the mountains and sea,
The rivers that curl....and the clouds that swirl.
Then the Earth gets small.....like a blue-green ball.

From your window you see....your own Planet Three.
And the Earth passes by ......as you look in the sky.
It gets so small......you can't see it at all.
Is it worth much at all.....this miniature ball?

Look at this planet.....how much is it worth?
Like a Christmas tree ball....so precious and small,
So tiny and blue....looking shiny and new.
Yet God chose this place.....one small planet in space.

For God so loved this Planet Earth,
That He gave us gifts of worth,
Gifts of water, sun and air,
Gifts of beauty everywhere.

God so loved this Planet Three,
That He gave to you and me,
Love that is so wide and deep,
Love that is for us to keep.

God loves "Earth people"...every one
So He sent His only Son.
Jesus brought God's love and grace,
To Planet Three in solar space.

Not to Jupiter or Mars,
Not to Venus or some stars.
Jesus lived on Planet Earth,
Christmas was His time of birth.

God loved us so much that He sent His only Son, Jesus, to this planet. Jesus came to tell us that God loves each and every one of us.

★ ★
★ ★
★ "For God loved the world so much that He gave ★
★ His only Son, so that anyone who believes in Him ★
★ should not perish, but have eternal life." John 3:16 (TLB) ★
★ ★

God shows His love in many ways.
Our planet is a treasure house of good things.
Can you think of some gifts
from God that you can see with your eyes?

God also gives us many invisible
gifts that we can't see with our eyes.
Can you think of some special
gifts from God that you can't see?

How would you finish these sentences?

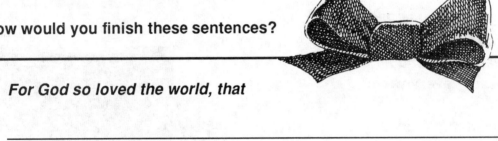

For God so loved the world, that

_____

For God so loves me, that

_____

Questions to Think About:

Take a penny and hold it near your eye. Hold it up to the moon.
How big does the moon look? How big does the penny look?
Why does the moon look like it is almost the size of the penny?
What is an orbit? Can a spaceship make an orbit around a planet?
Does the earth make an orbit? What does it go around?

Activities (See Activities Section)
Sky Diamonds (Earth/Sky) – p. 64
Sky Prayers – p. 65
Sky Box of Science Words – p. 63

View From Many Windows

When you look out of different kinds of windows, you can see different kinds of sky pictures. What can you see from a skyscraper window, a ship window, an airplane window, or a spaceship window? You can describe these sky pictures by writing or telling about them, or by drawing a picture.

Skyscraper Window
If you were up on the top floors of a skyscraper, your weather might be different from the weather down below on the city streets. How would a thunderstorm look from your window? Make a word picture, using words like skyscraper, skyline, and skyful of clouds. Or draw a window picture with the city and the storm outside.

Ship Window
What does the sky look like from a ship window, where you can see the sky and the ocean together? Make a word picture, using words like sky-blue, sea-blue, foamy white, and cloudy white. Or draw a picture of the sky and sea, using many different shades of blue.

Airplane Window
An airplane window is small, but you can see a big piece of sky. Sometimes you see cloud tops, and sometimes you are inside a cloud. Draw a picture or write about the inside and outside view of clouds from an airplane window. What would a faraway rainstorm look like? A nearby rainstorm?

Spaceship Window
Suppose that you are circling the earth in a spaceship, very high above the clouds. What kind of weather messages would be written by the clouds on our planet? Can you describe the light clouds racing the across the sea and land, the thick clouds lying like blankets, and the thunderstorm clouds? Make a word picture or draw a picture of your view from a spaceship window.

The Last View of the Earth's Sky

Suppose that you were chosen to be an astronaut who would travel to the stars. Before you could make the trip, you had to spend a year in a special city, getting training for your space voyage.

This special city was designed like a space station, so that people could learn how to live in outer space. There were no windows in the city, so you had to get used to artificial light, even in the daytime. Because you couldn't see the sky, the ceilings were painted blue. There was no weather because the city was closed off from the earth's atmosphere. There were no snowstorms, rain showers, or windy days. The temperature was always 70 degrees. Of course, you could always see the weather in the rest of the world by watching TV.

After a year of living in this city without windows, a giant lid was lifted up, and the city was open to the sky. This was done so that a space shuttle could come down and land inside the city. The shuttle would stay for a few days and then take you and other astronauts to the space station. For these few days, you could look up at the sky. This would be your last view of the earth's sky, except for a quick look as you traveled upward in the space shuttle.

What would you think as you watched the earth's sky for the last time?
Would you take pictures or write down ideas about it?
What would you remember about the sky?
Would you stay up all night to see the moon and the stars?
Would you enjoy a rainstorm or snowstorm for the last time?

How would you finish this story about the last view of the sky? Write down your ideas or tell about them.

ACTIVITIES

Take a cup and fill it up . . . Fill it up with fun!

Year-round projects can be done with water, snow, wind and sun.

1. ACTIVITY KITS 52
2. AIR EXPERIMENTS 53
3. BIBLE ACTIVITIES 54-55
4. CLOUDY DAY ACTIVITIES 56
5. EXPERIMENTS WITH LENSES 57
6. RAINBOW ACTIVITIES 58
7. RAINY DAY ACTIVITIES 59
8. SKY BOXES 60
9. SKY PICTURES 61
10. SKY WRITING 62-65
11. SNOWY DAY ACTIVITIES 66
12. STARRY NIGHT ACTIVITIES 67-69
13. STORMY DAY ACTIVITIES 70
14. SUNNY DAY ACTIVITIES 71
15. WEATHER ACTIVITIES 72-75
16. WINDY DAY ACTIVITIES 76
    REPRODUCIBLE PAGES 77-84

1. ACTIVITY KITS

It is fun for children to experiment with a wide variety of materials and use them in many different ways. Yet often it is hard for a busy teacher or parent to find time for such projects. Activity kits help children do projects on their own, with a minimum of help from busy adults. These kits are easy to put together, and fun to use. All you need is a box or bag, and some simple materials for a project.

Teachers always need an extra activity for the children who finish their work early and are bored. An activity kit could be the "surprise of the week" or the supplement for a science unit.

Kits can be "do-it-yourself" or "share-the-fun" projects. Grandparents or older children in the family might enjoy putting kits together, and helping a younger child. An activity kit could be left with the babysitter as an alternative to TV. Children can use activity kits with their friends and have fun in a creative way, rather than just hanging out.

Activity kits can be boxed as gifts. For example, the star kit could be put in a box or bag decorated with stars and tied with silver ribbon.

List of Activity Kits

Star Kit – Gummed stars, black paper, white paper, pencil, chalk, pictures of the constellations, shoe boxes, a nail and a flashlight. If you want to do stargazing, a telescope and binoculars would be useful . Use this kit for Star Activities.

Light Experiment Kit – One or two small mirrors, flashlight, hand lens, water glass, tape, scissors, old eye glass lenses, paper, balloon, cardboard, pins. This kit can be used for Sunny Day Activities, Rainbow Activities, and Experiments with Lenses.

Water Fun Kit – Tape, cards, paper plates, straws, poster paints, toothpicks, straws, wax paper, bowls, tea kettle, ice cubes, metal pie pan, and glasses. Use this kit for Rainy Day Activities.

Snow Activity Kit – Yardstick, coffee can, metal bread pan, hand lens, dark cloth, white paper, scissors, long rope, stop watch, and something to dress up snowmen. Use this kit for Snowy Day Activities

Weather Report Kit – Look up the Weather Activities for a list of materials. The basic things that you need are a thermometer, rain gauge, and notebook or chart to record your observations.

Travel Kit – Markers, crayons, colored paper, and pencils can be used to make sky pictures and sky writing in a notebook. A Frisbee or flying disk is fun to use.

Air Experiment Kit – Balloons, candle, candlesticks, jar, string, straws, tape, plastic cup, newpaper. Use this kit for Air Experiments.

2. AIR EXPERIMENTS

Balloon Rocket

MATERIALS: Balloon, straw, string, tape.
1. Cut a length of five feet of string. Thread it through a straw.
2. Tie the string to two chairs or other pieces of furniture. Or two children can hold the string. Blow up a balloon, and hold it shut with your hand.
3. Tape the side of the balloon to the straw, and slide the straw to one end of the string. Open your hand and let the balloon go. How fast does it go?
The push of the air as it goes out of the balloon is like the push of air as it goes out of a rocket. This is called "thrust".

Balloon Experiments

1. Blow up a balloon and hold the end together with your fingers. Hold it in front of you, and let go. What happens?
What direction does the balloon go? What direction does the air go?
2. Place a blown-up balloon on water, with the open end untied. Let go of the open end. What happens? **Does air move and push objects?**

Candle Experiments

MATERIALS : 1 jar, 2 candles and candlesticks.
1. Put the candles in candlesticks and light them. Put a jar over one candle. What happens? Did the candle go out? **Can fire burn without air?**
2. Light a candle, and put it behind a bottle or jar. Blow at the bottle. Does the candle go out? Why or why not? **Can air go around an object?**

Air and Water Experiments

MATERIALS: Plastic or paper cup, masking tape, newspaper or paper towel, pan, water.
1. Fill the pan with water.
2. Tape a crumpled piece of paper in the bottom of the cup.
3. Hold your cup upside down, and lower it straight into the water.
4. Bring it up again, and look at the inside of the cup. Is it dry? Did the paper get wet? Why or why not?
5. Now put the cup upside down into the water again. This time, tip the cup a little. Do you see bubbles coming out of the cup?
6. Take your cup out of the water, and look at the inside of the cup. Is it dry? Why do you think water entered the cup? What kind of bubbles came out of the cup?
Can air and water be in the same place at the same time?

■ ACTIVITIES / 3. BIBLE ACTIVITIES

3. BIBLE ACTIVITIES

There are many ways to help children discover the treasures in the Bible. For example, Bible storm stories can be dramatized with different sounds for the wind, rain, and thunder. Storm stories can also be read on stormy nights when the sound effects are outside. Star verses can be read by flashlight outside on a starry night. The Christmas story can be read by candlelight with all lights turned off.

Bible verses in this book can be used as handwriting exercises. They can be copied and illustrated with small sky pictures, such as a border of stars, a border of clouds/rain/lightning, etc. Bible verses about the sky can be written inside a kite, a balloon, or rainbow.

Rainbow Bible Bookmarks

A rainbow of ribbons in a Bible reminds us that the Bible is full of treasures to discover. Ribbons of many colors can be used as Bible bookmarks. Here are some ideas for using ribbon bookmarks for Bible verses in this book, and favorite passages of your own choosing.

Different colors of ribbons can represent different Bible topics .
  Yellow ribbons for Bibles verses about stars, sun and light.
  Blue ribbons for storm stories, and verses about air, sky, water.
  Red ribbons for verses about God's love and care for us.
  White ribbons for verses about God and Jesus.
  Other colors for other topics of your own choosing.

It is a good idea to keep the ribbons in the Bible from previous readings, so that there is a colorful rainbow of ribbons in the Bible. This makes it easier to find a favorite verse or story.

You could tape a small circle of paper on the end of each ribbon, and draw a small symbol, such as a star, to represent the story or verse. For example, the storm story in Mark 4:35-41, could have a blue ribbon with a small boat picture.

Underlining favorite verses in different colors is another way to use the color concept. Underlining and ribbon bookmarks help children locate favorite verses quickly.

Seek and Find games can be played, using ribbons and other color cues. Give out ribbon bookmarks, and see how quickly the game players could find the verses, and put a ribbon of the right color in the right place.

For beginning readers, it may be enough to locate the passage, and let a parent, older brother, or sister read it. Even a young child who can't read, can find a passage in the Bible which has already been marked with a ribbon bookmark.

Bible Stories and Verses

Verses about Water, Air, and Sky (blue ribbon bookmark)
Clouds – Job 36: 26-29, Psalm 147:8
Storms – Job 37: 5b-7, 9-11, 14,15,16,18,
Questions about the Sky – Job 38:19, 22, 28, 32, 35
Wind – Psalm 147:18   Water – Psalm 65:9a

Storm Stories (blue ribbon bookmark)
Forty Days of Rain, an Ark and a Rainbow – Genesis 7, 8.
A Storm at Sea, a Whale and Jonah – Jonah I and 2.
Thunder, Lightning and the Power of Storms – Job 37:3-16
A Storm, a Boat, and Jesus – Mark 4:35-41 or Luke 8:22-25
House Built on a Rock – Matthew 7:24-27
A Big Storm and Shipwreck – Acts 27:13-44.

Star Verses (yellow ribbon bookmark)
God Created Stars – Genesis 1:16, Job 38:7
Praise the Lord, O heavens! – Psalm 148:1, 3, 4.
Counting Stars – Psalm 147:4
Star of Bethlehem – Matthew 2:2-12
Morning Star – Revelation 22:16

Light Verses (yellow ribbon bookmark)
A Great Light – Isaiah 9:2
God Created Sun, Moon, Stars – Genesis 1:16, Psalm 75:16
Light of the World – John 8:12
Light for Blind Men – Matthew 20:29-34
Looking Through a Glass – I Cor. 13:12

Verses about God and Jesus (white ribbon bookmark)
God's Power – Psalm 62:11b
God's Work of Art – Ephesians 2:10a
God Answers Prayer – Jeremiah 33:3
Jesus Christ Forever – Hebrews 13:8
Jesus With Us – Matthew 28:20b

Love Verses (red ribbon bookmark)
God is Love – I John 4: 7, 8, 12
God's Love Forever – Jeremiah 31:3
Love One Another – Leviticus 19:18, Matthew 19:19
God Loves the World – John 3:16

4. CLOUDY DAY ACTIVITIES

Cloud Shapes

Have you ever seen funny cloud shapes in the sky — cloud mountains, animals, castles, giants and funny faces? Have you ever seen feather clouds or mashed potato clouds? Look for funny cloud shapes whenever you are in the car or just looking out the window.

It is fun to make drawings of interesting cloud shapes. And you can also write sentences about the funny clouds, if you want. *A cloud giant marched across the sky in giant sky boots. I saw a cloud castle with mountains all around it.*

Recipe for Clouds

Can you make a very small cloud in your kitchen ?
MATERIALS: *Jar, boiling water, ice cubes, string or rubber band, thin cloth.*
1. Boil water on the stove. What happens when the water boils?
2. You see steam coming out of the boiling water. Steam is condensed water vapor.
3. Pour the hot water into the jar. When the jar is hot, pour out all but one inch of the water. Put the cloth over the jar, and secure it with a string or rubber band. Put the ice on the top of the cloth. What happens when the hot air from the water meets the cool air from the ice? Do you see a cloud in your jar?

Cloud Watching

Clouds seem brighter if you look at them upside down. Lie on your back and watch them for a few minutes. Or look at clouds with binoculars or a telescope.

Here is a project for a cloudy, windy day. Take a medium-sized mirror, a crayon, and a pocket compass outside. Place the mirror on the ground so that it is in line with the compass points. Mark the compass points with a crayon on the mirror using the letters, "N,S,E,W" for north, south, east, and west. Stand over the mirror and watch the moving clouds. Which way are they moving? *(Don't stare into the sun's reflection in the mirror.)*

Cloud Mobile

Cut out different shapes of clouds, using light cardboard or heavy paper. You could also cut out stars, planets, sun, and moon. Hang on a coat hangar, using thread or string.

5. EXPERIMENTS WITH LENSES

Experiments with a Hand Lens

1. Focusing with lens – Hold the lens about four inches from your eye. Look at an object which is about six inches from your eye. Move the lens up and down, until the object is clear.
2. Snowflakes – Catch a snowflake and put it on a cold piece of cloth (which has been in the freezer). Look at the snowflake with the hand lens and describe it or draw it.
3. Frost crystals – Scrap frost from a window or refrigerator, and look at it under the lens.
4. Other crystals – Look at salt and sugar crystals with the hand lens. Draw them, if you want to.
5. Other ideas – Look at your fingerprints, the spices in the kitchen spice rack, a newspaper, a postage stamp.

Experiments with other lenses

1. Water can act as a lens, and make things look larger. Cut a small one inch hole in an index card. Place a piece of tape across the hole. Put a drop of water on the tape, using a straw or eye dropper. Look at some printing on a page. Do the letters look larger?
2. A water glass can be a lens. Look through the bottom of a thick glass at a newspaper headline. Move the glass up and down to change the size of the print. Does it look larger or smaller?
3. Try looking through someone's glasses. Do they make things look bigger or smaller? Convex lenses make things look bigger. They are thicker in the middle, and curve outward. Concave lenses make things look smaller. They are thinner in the middle and thicker at the edges.

Experiments with Binoculars

1. Focus the binoculars so that you can see clearly. Look through the binoculars at some object. Close one eye and focus the lens for the other eye. Then repeat the process for the other eye. Now open both eyes, and make sure you can see clearly.
2. Pick out something which you can see from a window, but not clearly. Look at clouds, birds, planes, trees, faraway buildings or mountains. Pick out details, such as lettering on a sign, colors of a bird, or leaves on a tree.
3. Pick out a special place to study with your binoculars. Every day for a week or more, you can use the binoculars to look at this place. Write down or talk about the details that you see.

6. RAINBOW ACTIVITIES

Indoor Rainbows

MATERIALS: *Pocket mirror, sunny window, a glass, and a prism (optional).*
1. Put the mirror in a glass of water.
2. Place it in a sunny window.
3. Keep changing the direction of the mirror until it catches the sunlight and reflects a rainbow on the wall.
4. You can also use a prism to make a rainbow. Place the prism in a sunny window so that it catches the sunlight and reflects a rainbow on the wall. Keep moving it to change the shape of the rainbow.

Outdoor Rainbow

Go outside on a sunny day, and stand so the sun is shining on your back. The sun should be halfway up the sky, not overhead. Turn on the hose, and make the water come out as a fine spray by adjusting the spigot. Wave the hose around and watch for a rainbow to appear in the mist of the hose spray. You can also change the hose nozzle to make the waterdrops bigger or smaller. Does that change the rainbow?

Rainbow Poems and Pictures

Make a rainbow on a lined piece of paper, and then write a poem inside the rainbow. The poem should have one word on the first line, two words on the second line, and three words on the third line, etc.

Place a prism so that the rainbow falls on a piece of paper or the wall. Make a drawing of the prism rainbow.

Make a rainbow with crayons, then paint over it with blue paint for sky color. The rainbow will still show through the paint if you use thin watercolor paint. Be sure to add water to the paint if it is too thick.

7. RAINY DAY ACTIVITIES

Raindrop Picture

Can you make a picture with raindrops and paint?
MATERIALS: Poster paint, or food coloring, paper plate, and raindrops.
1. On a rainy day, splatter drops of paint or food coloring on a paper plate, but don't mix the colors.
2. Hold the paper plate out the window in the rain. It may take only half a minute for the rain to mix up the colors.
3. When you like the mixture of raindrops and color drops, bring the plate inside to dry.

Waterdrop Fun

A waterdrop always forms a round shape.
MATERIALS: Wax paper, toothpicks, and water.
1. Sprinkle some waterdrops on waxed paper. Does the water always form round drops? Break the drop into smaller and smaller drops, using the toothpick. Are each of the small drops round?

Waterdrop Game

Can you drip a drop?
MATERIALS: Straws, water in a glass, and bowls. You will need two or more people to play this game.
1. Fill each glass with water, and put the straw in the glass. Each person should hold their finger over the top of the straw.
2. Lift the straw out of the glass and place it over the bowl, keeping a finger on the top of the straw. Now move the finger aside just slightly to make the straw drip a single waterdrop.
3. The object of the game is to see who can release the most single drops. This is harder than you think. When you take your finger away from the straw, air rushes in the top and pushes the water out the bottom.

Recipe for Rain

Can you make rain in your kitchen?
MATERIALS: Tea kettle, ice cubes, metal pie pan, water.
1. Boil water in the tea kettle, until there is steam coming out of the spout. The steam is clear, and just beyond the steam, a "cloud" will form.
2. When the "cloud" forms, hold the pie pan with the ice cubes in the cloud. What happens when the "cloud" hits the icy cold pie pan? Water droplets will form on the pan. These droplets are like cloud droplets in the sky. Cloud droplets are made when water vapor cools off and condenses.

8. SKY BOXES

What would you put in a sky box? At first that sounds like a silly question. Of course you can't put the big sky in a small box. But there are some other things you can put in sky boxes.

Sky Activity Boxes

Find a large box and put in basic materials for the sky experiments in this book. You can also make up your own experiments, using these materials.

Science Materials – flashlight, hand lens, a pocket mirror, and a larger one, telescope, binoculars, star map, ruler, or yardstick. You can also put in items for weather projects. This might include a thermometer, rain gauge, barometer, wind vane, or whatever you have on hand.

Household Materials – coffee can, candle and candlesticks, shoe boxes, glasses and jars, pans, straws, plastic wrap, rubber band, string, tape, toothpicks, balloon, cloth, nail, paper plates, eye dropper, scissors, pins, cardboard.

Art and Writing Materials – notebook, drawing book, white or colored paper, crayons, markers, pencils, poster paint, chalk, gummed stars.

It is useful, but not necessary to make a sky activity box. These materials can be collected in one large box, or put in activity kits for specific activities. Or you can simply get the materials when you do the project.

Sky Treasure Box

You can't put a rainbow, or a snowflake or a beautiful sunset in a box. But you can paint a rainbow, cut a paper snowflake, or take a photo of a sunset, and put these beautiful pictures in your box. And you can put in sky poems, stories, and sky verses from the Bible.

Sky treasures, such as pictures, can be put up on the wall, or the refrigerator door. It is also a good idea to send sky pictures or sky poems to sick people or shut-ins, who can't get outside to enjoy the sky.

9. SKY PICTURES

How does the sky change day by day? Draw sky pictures in a notebook or on paper and see how each picture is different.

MATERIALS: Notebook or drawing pad, white or colored paper, crayons, markers, pencil, chalk, gummed stars, watercolor paint.

DIRECTIONS:

1. On a cloudy day, draw clouds on blue paper with chalk. Or draw cloud shapes on white paper, and color around them with a blue color.
2. Use crayons or markers to make a sunset, sunrise or rainbow. Be sure to put in the blue sky color, too.
3. On a stormy day, make a gray sky with clouds, lightning and raindrops. Use crayons, markers or chalk.
4. Make a night picture with black paper and gummed stars. You can draw patterns of the constellations with white chalk.
5. Draw the stars and moon on white paper, using yellow crayon. Paint over the whole picture with black paint. The yellow stars and moon will show through the black paint.
6. Use your imagination and many colors to make a sky filled with colorful things – kites, balloons, rainbows, birds, and all.
7. Make a sunrise picture using the colors blue, red, yellow, and orange. Use water colors, crayons or markers.
8. If you are using this book as a workbook, fill the pages with sky pictures – rainbows, balloons, kites, clouds, birds, airplanes, etc.
9. Draw a window frame. Fill it with a sky picture that you can see from your window. Or draw the picture that you could see from a skyscraper, a ship's window, an airplane window, or a spaceship window. Read "View From Many Windows", to get some ideas.
10. Take photographs of a sunrise, sunset, clouds, jet trails, etc.
11. Color the pictures on pages 79, 81, and 82, and add your own drawings or writings to the pages.
12. Gather your pictures together in a decorated folder or a sky box. Show your sky pictures to your family, friends, and schoolmates.

■ ACTIVITIES / 10. SKY WRITING

10. SKY WRITING

Sky Notebook

A sky notebook can have anything about the sky that you want to include. Here are some ideas:

Sky Pictures: *magazine pictures, photos, drawings and paintings.*
Sky Writing: *poems, stories, jokes, skygrams and questions about the sky.*
Weather reports: *pictures or words to tell about each day's weather.*

Sky Writing and Discussion

1. Write a letter describing the sky to someone who's never seen it. Or have someone shut their eyes while you describe the sky to them.

2. Write, draw, or paint a thank-you letter to God for the sky.

3. Make up a TV commercial selling the sky.

4. Write a sky-gram from Planet Earth to another planet. Draw small sky pictures around the edges of the page.

5. Tell or write about your own sky wonders and sky questions that you have thought about. Read again "A Cup of Wonder" to get some ideas.

6. Where does the sky end? Give your own answer to this question.

7. Make up a sky box of words on a subject such as snow, and then write a poem or paragraph.

8. Draw a large sky shape, such as a cloud, a star, a kite, or a balloon. Write a poem, prayer, or collection of words inside your sky shape.

9. Draw a rainbow on a lined piece of paper, and write a poem inside the rainbow. The poem should have one word on the first line, two on the second line, etc. Or use the rainbow on page 79 and write words inside it.

10. Write down your own ideas about the readings, "View From Many Windows", and "The Last View of the Earth's Sky".

ACTIVITIES / 10. SKY WRITING

Sky Word Box

Here is a boxful of sky words, which can be used for poems, jokes, riddles, and stories. You can make up your own sky words.

Colors and light
sky-blue, sky gray, sky glow, skylight (*window in the roof*).

City words
skyscraper, skyline (*city against the sky*).

Other sky words
sky map(*star map*), skywriting, skyway, sky cover*(clouds)*, skyful, skyrocket.

Star Words
starbright, stardust, stargazing, starful, starlight, starless, star map, star shell, starship.

Sun Words
sunbeam, sundown, sunflower, sunlight, sunrise, sunset, sunshine, sunup.

Sky Box of Science Words

You can fill a sky box with words about stars, storms, weather forecasting, or any other science subject. These words can be used for writing, discussion, or dictionary work. Here are some words about the weather.

air molecules	condensation	flood	stratus cloud
atmosphere	constellation	fog	temperature
barometer	crystals	freezing point	thermometer
boiling point	cumulus cloud	frost	water vapor
Celsius	dew	hail	weather satellite
cirrus cloud	evaporate	humidity	weather station
	Fahrenheit	rain gauge	

63

Sky Diamonds

The beauty of the sky is ours, not just for a few days, but every day. We can always look up at the sky and enjoy it because it's a free gift from God.

How can we thank God for the beautiful sky and everything in it? One way is to make a sky prayer. A sky diamond is thank-you prayer for sky wonders.

First, make a list of words that you want to put in your sky diamond, using one of these topics:

(1) **Visible/Invisible** – *See illustrated example below.*
(2) **Our Sky/Outer Space** – *Example: Our sky – air, water, sun, light, blue color.. Outer space – no air, sun, planets, black color.*
(3) **Earth/Sky** – *Example: Earth – rivers, mountains, oceans, etc. Sky – stars, meteors, planets, etc.*
(4) **Other Topics** – *Big/Little, Clouds/Sun, Snow/Rain.*

When you have finished your list, fill in the sky diamond with the things for which you are thankful. Fill the corners with tiny illustrations, if you want. Use the words in your sky diamond for a sky prayer.

You can also fill a sky diamond with words which describe one subject, like snowflakes or stars. The first line has one word *(the subject)*, the second line has two descriptive words, third line has three descriptive words, fourth line-two words, fifth line-one word. *See example below.*

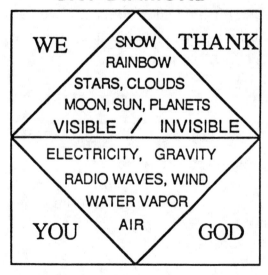

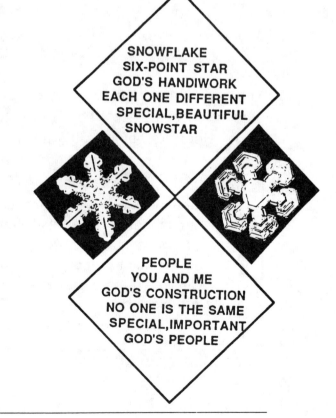

Sky Prayers

Wonder about the sky can be expressed in sky prayers. Litanies are one way to do this. A litany has two parts, one for the leader, and another for the group. For example:

Leader: For sky colors—blue, orange, yellow, and red.
Group: God, we thank you.
Leader: For flying things—birds, airplanes, and kites.
Group: God, we thank you.

Responses for litanies include: Thanks be to God, Thank You, God, etc.

Another way to express prayers is by writing prayers inside different shapes. A prayer can be written inside a rainbow, cloud, sun, star, planet, or any other shape. Another idea is to draw a sky picture and write the prayer inside the picture frame. Both ideas are illustrated below.

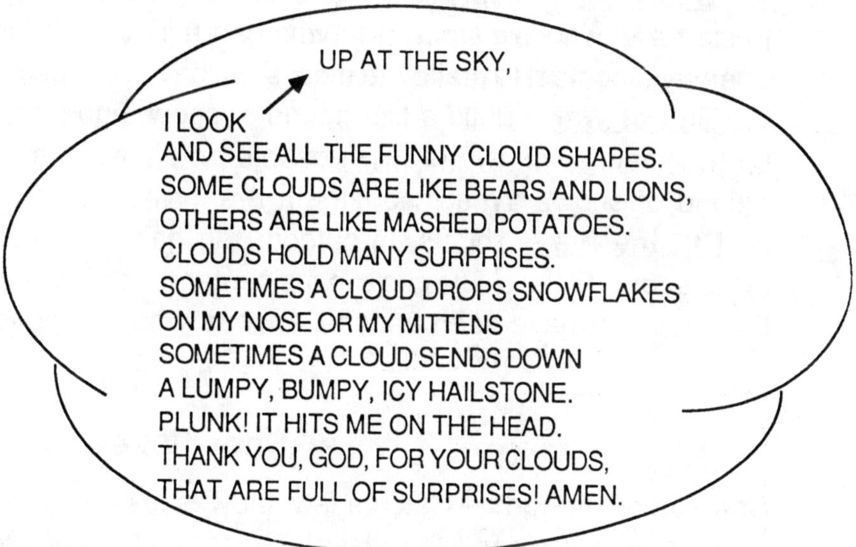

I LOOK UP AT THE SKY,
AND SEE ALL THE FUNNY CLOUD SHAPES.
SOME CLOUDS ARE LIKE BEARS AND LIONS,
OTHERS ARE LIKE MASHED POTATOES.
CLOUDS HOLD MANY SURPRISES.
SOMETIMES A CLOUD DROPS SNOWFLAKES
ON MY NOSE OR MY MITTENS
SOMETIMES A CLOUD SENDS DOWN
A LUMPY, BUMPY, ICY HAILSTONE.
PLUNK! IT HITS ME ON THE HEAD.
THANK YOU, GOD, FOR YOUR CLOUDS,
THAT ARE FULL OF SURPRISES! AMEN.

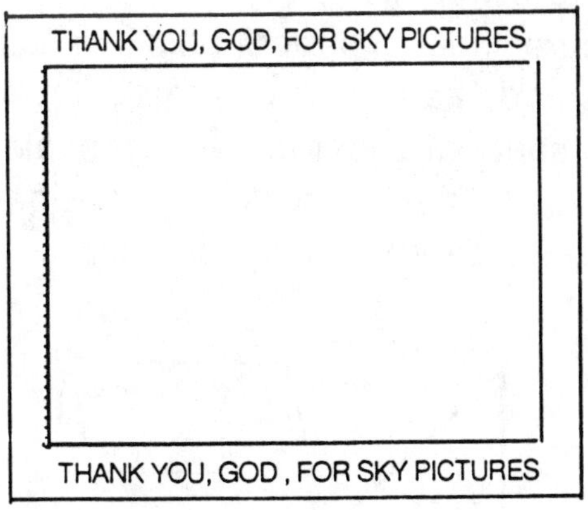

THANK YOU, GOD, FOR SKY PICTURES

THANK YOU, GOD, FOR SKY PICTURES

SUNRISE
OPEN YOUR EYES
TO SEE GOD'S SURPRISE
IN THE MORNING SKIES

11. SNOWY DAY ACTIVITIES

Snow Shapes

MATERIALS: Lots of fresh snow, a metal bread pan, tree branches, rocks, and materials from your closet.

1. **Snow Bricks** can be made by using a bread pan as a mold. Pack the snow in the pan, then turn it over and tap it lightly to remove the brick. If the snow is light and powdery, sprinkle a little water on it as you pack it. Snow bricks are good for making snow forts, walls and igloos.
2. **Snow monsters** are fun to make. Collect big mounds of snow, and shape them into monsters, such as a giant snake, shark or spider. Use branches for legs and rocks for teeth. Make your monster as scary as possible.
3. **Snowmen** can be decorated with funny hats and bright scarves, and sun glasses.

Snow Games

MATERIALS: Lots of snow, a long rope, and a stop watch or timer.

1. **Snowball contest** – Have one person hold the stopwatch, and tell people when to start and when to stop. See who can make the biggest snowball in one minute. Then try a contest to make the most snowballs in three minutes. Make up other ideas.
2. **Tug of War** – Build a big mound of snow, and stretch a long rope over it. Make up two teams with two or more people on each team. Try to pull the rope over the mound of snow. What happens to the losing team?
3. **Cut the Pie** – This is an outdoor tag game. Trace a large wheel with spokes in the snow. Make a little circle in the center which is the safe place. Play tag, following the lines that have been traced. Anyone who is tagged or runs outside the lines is "it".

Snowflakes

Snowflake Designs – Catch a few snowflakes on your mittens or a dark cloth. Look at them with a magnifying glass. Do they all have six points?

Paper Snowflakes – Six-pointed: Start with a square sheet of paper. Fold it in half. Then fold it in thirds. *(See diagram)* Cut notches and shapes in the edges. Open it up to see design. Eight-pointed: Fold a square sheet of paper in half diagonally. Fold in half again. And once more. Round off points to make curves. Cut designs in the folded edges. Open it up. Snowflakes make nice window decorations, and they can also be glued on colored paper.

6-pointed

8-pointed

12. STARRY NIGHT ACTIVITIES

Family Fun Under the Stars

★ Have an evening picnic in your backyard or a park and watch the stars come out. Or sleep under the stars in tents and sleeping bags in your backyard or a campground.

★ Stop the car when you are driving on a starry night and take a few minutes to look at the stars and find constellations. Have a whole family hug before you get back into the car.

Star Show Activity

Can you make a star show with a flashlight and boxes?
MATERIALS: Cardboard box (such as a shoebox or cereal box), flashlight, and nail. Pictures of star constellations, such as the Big Dipper.

1. Punch holes in the bottom of the box to make a pattern of stars. For example, you could make the Big Dipper pattern.
2. Put a lighted flashlight inside the box, so the light shines out the holes.
3. Turn out the lights, and place your star box so that the light reflects on a wall. You will see the constellation pattern as dots of light reflected on the wall.
4. You can make several of these star boxes and then have a star show.

Star Pictures

Can you make pictures of the constellations?
MATERIALS: Black paper, white paper, pencil or chalk, tape, gummed stars, and pictures of constellations.

1. Draw dots for star patterns with chalk or pencil on the black paper.
2. Put gummed stars on the dots to make star pictures.
3. Make white labels with black lettering to tape on your pictures.

Pie in the Sky

How many stars can you count in the sky?
MATERIALS: Flashlights, twenty or thirty small stones, clear night sky.

1. Go outside on a clear night, and try counting stars.
2. Divide the sky into four or more sections, just as you would divide a pie into pieces. Let each person count one section or more.
3. Use a flashlight as a pointer, and small stones as counters—one for each one hundred stars. Who counted the most stars? How many did everyone count? You should be able to count more than 2,000 stars on a very clear night.

Star Verses in the Bible

There are many verses about the sky and the stars in the Bible. You can read them by flashlight outside on a starry night. Or you can look them up when you are inside.

> "Praise the Lord, O heavens!
> Praise him from the skies...
> Praise him, sun and moon,
> and all you twinkling stars."
> Psalm 148:1,3. (TLB)

You can make up some new verses by adding space-age words. Here are some words you might use:

★ ★

universe	*cosmic space*	*solar system*	*galaxy*
planetoids	*dwarf stars*	*asteroids*	*comets*
giant stars	*supernova*	*radio stars*	*meteors*

★ ★

Fill in the blanks in this verse with space-age words:

> Praise God from the reaches of outer space.
> Praise Him all planets,
> Praise Him all _____ and _____.
> Praise Him all _____ and _____.
> Praise Him all _____ and _____.

★ ★

Other Star Verses

God Created Stars – Genesis 1:16, Job 38:7, Psalm 8:1-4.
Counting Stars – Psalm 147:4
Star of Bethlehem – Matthew 2:2-12
Morning Star – Revelation 22:16

If you read the Christmas story (Matthew 2:2-12), you could make a star show, using the Star Show Activity in this section.

Special Star Activities for Schools

1. Make nighttime sky assignments for children. They can report on locating the North Star first, then find other constellations. The Big Dipper is the easiest to find all year long.
2. Plan a field trip to a local planetarium or observatory.
3. Star watching in the evening could be planned as a joint parent-teacher project.
4. Learn the meaning of these words: star, planet, constellation, and galaxy. Use a flashlight at night to point out each of these sky wonders.
5. Learn the meaning of the word "infinity". Use a flashlight to show the limitless expanse of sky. The light does not bounce back, as it does from the ground, it just goes on and on.

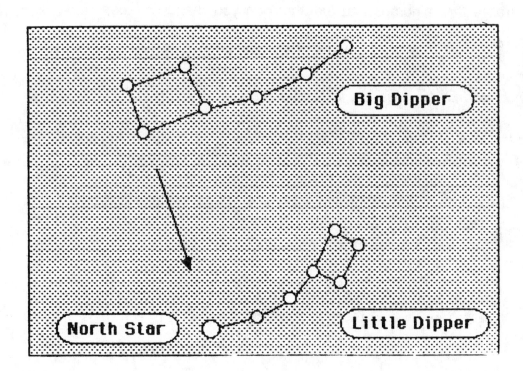

The North Star is always in the same place in the sky. You can see it if you live in the northern half of the earth. In Canada, the North Star is higher than it is in Florida or California. You can find the North Star by looking for the pointer stars in the Big Dipper. *(See the picture above.)* **The Big Dipper and Little Dipper and all the other stars, rotate around the North Star. But the pointer stars always point to the North Star. This picture is a spring picture. In the fall, the Big Dipper is below the North Star, and may be so low that you cannot see it.**

13. STORMY DAY ACTIVITIES

Storm Experience

During a storm, stand out on the porch, or in your open garage, and experience the storm, using all five senses.

1. **Listen** to the soft drumbeats of the rain and the booming drumbeat of thunder. Can you hear the up-and-down melody of the wind?
2. **Feel** the wind and the coolness of the air.
3. **Watch** the changing clouds and the raindrops dancing on the street. What changes take place when a storm comes in? Take some photos of the storm changes, if you want to.
4. **Smell** the air, which has been washed clean by the rain.
5. **Taste** the raindrops and feel them on your hand or face.

Storm Stories in the Bible

On a stormy day, it is fun to read a Bible storm story.

Forty Days of Rain, An Ark and a Rainbow – Genesis 7 and 8.
A Storm at Sea, a Whale and Jonah – Jonah 1 and 2.
Thunder, Lightning ,and the Power of Storms – Job 37: 3-16
A Storm on the Sea of Galilee – Mark 4: 35-41 or Luke 8: 22-25.
House Built on a Rock – Matthew 7: 24-27
A Big Storm and Shipwreck – Acts 27:13-44.

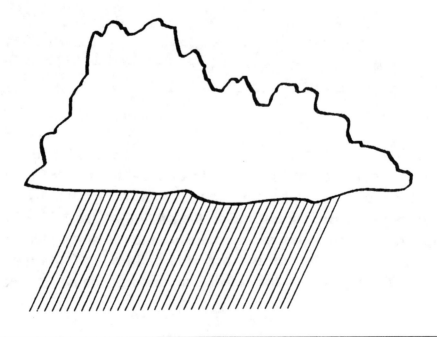

14. SUNNY DAY ACTIVITIES

Sun Tricks

MATERIALS: *A hand lens, bright sun, and a piece of paper and balloon.*
1. Put a piece of paper on the ground, and hold the hand lens over it so that the sun's rays shine through the glass and onto the paper. Make the lens move up and down until you get a very small and very bright dot of sunlight on the paper. How long do you hold the lens before the paper catches fire?
2. Blow up a balloon and tie it shut. Then tape it to a fence or outdoor table. Hold your hand lens so that you get a small bright dot of light on the balloon. How long does it take before the balloon bursts?

Sun Tag

MATERIALS: *One pocket mirror for each person who plays, sunlight, and a wall. It is best if the wall has some shadows on it.*
1. Hold your mirror so that it reflects a spot of light on the wall.
2. Practice moving your mirror, so that your sunspot moves.
3. Choose one child to be "IT". "IT" chases the other sunspots around on the wall until it touches another spot. Then the other spot becomes "IT".

Shadow Tag

You need three or four players, and a sunny afternoon to play this game. The person who is "IT" chases the other players around, trying to step on their shadows. When "IT" steps on a person's shadow, "IT" yells "Stop!" Everyone stops while "IT" counts to five. Then the person whose shadow is stepped on, is the new "IT", and starts chasing the other players.

Sun Picture

Make a sunrise picture using the colors blue, red, yellow, and orange. Use water colors, crayons or markers. Or get up early and take a photo of the sunrise. Write a sentence prayer under the sunrise, thanking God for the sunrays, sunbeams, sunflowers, and sun power.

Sun and Shadows

Stand in the sun and have someone measure your shadow. Do this early in the morning, at lunchtime, and late in the afternoon. When was your shadow the longest, the shortest? Why?

Draw a big circle on a piece of cardboard. Put straight pins around the edge of the circle. Shine a flashlight on the pins, and find the shadow of the pins. Now move the flashlight so that the shadows of all the pins fall in the same direction. Make all the shadows fall outside the circle. Do the shadows fall toward the flashlight or away from it?

15. WEATHER ACTIVITIES

Quickie Weather Report

In the morning, look at clouds in the sky, and the outdoor thermometer. Open the window to feel the air and listen for wind. Make a morning weather prediction to your family. At dinnertime, talk about the day's weather. Was your early weather report correct?

Newspaper/TV Weather Report

Look at the newspaper or TV weather report every day for a week. Each day, compare yesterday's weather report and the actual weather that you experienced. Why are there differences between the forecasted weather and the actual weather?

Weather Chart

It is fun to keep track of the daily weather. There is a weather chart in the back of this book for you to use. Every day you can look at the sky and write down the names of clouds you see. If you have a rain gauge or yardstick, you can write down the amount of rain or snow. Write down the temperature and a very brief description of the weather. Or draw a small weather picture. Estimate the wind speed, using the following chart:

WIND SPEED CHART

Miles Per Hour

0	Smoke rises straight up.
1-3	Direction of wind is shown by smoke drift.
4-7	Wind felt on face, leaves rustle.
8-12	Leaves and twigs in constant motion.
13-18	Small branches move, dust and papers are raised.
19-24	Small trees begin to sway.
25-31	Large branches move, telephone wires whistle.
32-38	Large trees move, it is hard to walk against wind.
39-46	Small tree branches break off.
47-54	Roof shingles blown free, buildings damaged.
55-63	Trees uprooted, buildings damaged.
64 +	Widespread damage, very rarely experienced.

Thermometer Activities

1. Take the outside temperature readings every day at the same time and place for a week. Be sure to pick a shady place for your thermomenter, so that the sun's direct heat doesn't make the reading too high. Write the temperatures on a chart. Check to see if your daily temperatures are the same as the temperatures in the daily weather news.
2. Record the temperature at different times and places outside for a week. Place the thermometer in sunny and shady areas, and keep it in the same place for at least ten minutes or more before moving it to another place. Make a chart of the temperature readings. Talk about the reason for the differences in temperatures.
3. Take inside and outside temperatures, and compare them.
4. Find the average temperature for your area for a week or a month. To do this, you need to keep a list of the daily temperatures. You can use the temperature numbers in the newspaper, or take your own temperature readings. You can write down the evening or morning temperatures, or both.
5. Make a bowl with cold water and ice cubes. Put your thermomenter in the bowl for several minutes. It should register 32 degrees *(0 degrees C)*.

Measuring Rainfall

To measure rainfall, you need a rain gauge. You can buy or make one. A rain gauge should be placed outside away from trees and buildings, on a level spot. If you have a plastic gauge, bring it inside if the temperature is below 32 degrees F. Otherwise, it might crack if the water freezes.

To make a rain gauge, you will need a small jar, a ruler, a permanent black marker, and a can *(such as a one pound coffee can)*. The jar will be used to measure the rainfall, and it should have a small diameter *(2-3 inches)*. The can will be used to catch the rainfall, and it should have a larger diameter *(about 5 inches)*. A funnel would be useful for pouring the water from the can into the jar.

1. This is how you mark the jar. Pour water into the can until it is one inch deep. *(Measure with your ruler.)* Now pour this water into the jar.
2. Mark the water level on the outside of the jar with a black line, and write 1". Now mark a line halfway between the 1 inch mark and the bottom of the jar. Write 1/2 on the line. Then mark a line halfway between the 1/2 inch mark and the bottom of the jar. Write 1/4 on the line.
3. This is how you use the rain gauge. Put the can in a level open space, and fix it so it won't tip over. After each rainstorm, bring the can inside and pour the rain water into the marked jar, so that you can measure it.

Measuring Snowfall

To measure snow, you need a yardstick or marked wooden rod. The rod can be marked in half inch increments. Take the yardstick or rod outside and measure the snow in different areas. How deep is the snow next to the road or a building, a field or a yard? How much deeper are the snowdrifts than the rest of the snowfall?

How many inches of snow makes one inch of water? You can fill a coffee can with unpacked snow, and measure it with a ruler. Then bring it inside and let it melt. Measure the melted water with a ruler. How many inches of snow make one inch of water? Usually it takes twelve inches of snow to make one inch of water.

Kinds of Clouds

No two clouds are alike, but there are three basic types- cumulus, cirrus, and stratus.

Cumulus clouds are puffy with flat bases and round tops. Their shapes change all the time, and they have sharp outlines. They form in the rising warm air in the daytime, and disappear at night.

Cirrus clouds are thin and feathery. They are very high in the sky, about 25,000 feet or more. Because the air is very cold in the high regions, these clouds are made up of ice crystals.

Stratus clouds are dull gray, usually low, and layered. They cover most of the sky.

There are also combination types of clouds. We put two words together to describe them.

Cirrostratus clouds are thin sheets of cirrus clouds which are spread out over the sky, and are very high.

Cirrocumulus clouds make a thin, wavy pattern high in the sky.

Cumulonimbus clouds are often called thunderheads. They have flat bottoms, and high, towering tops, which get darker as the storm comes in. You see these clouds most often in the summer. (*Nimbus means rain.*)

Nimbostratus clouds are thick and dark gray. They are low in the sky, and very wet-looking. You can see them any time of the year.

Altostratus are high, spread-out sheets of clouds. (*Alto means high.*)

Altocumulus are high, puffy clouds.

Usually, the rain clouds are the low, dark, heavy ones. Cumulus clouds can start out as small, light, and white ones, but as they get larger, darker, and heavier, the chance of rain increases. Cirrus clouds are very high and light, so they are not rain clouds. Stratus clouds may bring rain, if they are dark, heavy, and low.

Weather Chart

(Each day you can record the temperature, wind speed, amount of rain or snow, cloud shapes, and anything else you want to say about the weather.)

Date	Temp	Wind	Rain	Cloud Shapes	Snow	Description of Weather

16. WINDY DAY ACTIVITIES

Flying Discs

Can you make something to catch the wind and fly?
MATERIALS: Plastic plates, metal pie pans, paper plates, or cardboard circles. Make several flying discs of different materials, and try them out.

1. Which disc will go the farthest? How many feet will it go? Measure the distance with a yardstick.
2. Which is the best way to throw the disc? Try overhand, underhand, waist high and chest high.
3. Can you make your discs fly toward a target? Try flying the discs at targets that are ten feet away, fifteen feet away, etc.

What two kinds of power did you use? Aren't you glad that God made wind power and muscle power?

Parachutes

Can you make the wind work for you?
MATERIALS
1. One handkerchief, square of cloth, or paper napkin.
2. One weight, which can be a clothespin, piece of wood, etc.
3. String or thread, and tape.

DIRECTIONS
1. Cut four pieces of string or thread that are 12 inches long.
2. Tie or tape the pieces of string or thread to the four corners of the material. Tie or tape the four threads or strings to the weight.
3. Fold up your parachute and toss it out a window or high up in the air. Why does it come down so slowly?

Parachute Tests

Try different types of parachutes, and see which one comes down more slowly. Here are some ideas:
★ Larger parachutes.
★ Longer strings.
★ A hole in the top of the parachute.
★ A circle parachute.
★ A plastic parachute.

Try different kinds of objects, and see whether the heavier or lighter ones come down faster.
You could try a crumpled ball of paper, a cork, a film can with pennies or rocks in it, or a small plastic figure.

REPRODUCIBLE ■

Cup of Wonder

Write about your own sky wonders and sky questions.

Sky Box

Fill this window frame with a sky picture that you can see from your window, or draw an imaginary scene.

Rainbow Picture

Write a poem or prayer inside the rainbow.
Use paints, crayons or markers to color the rainbow.

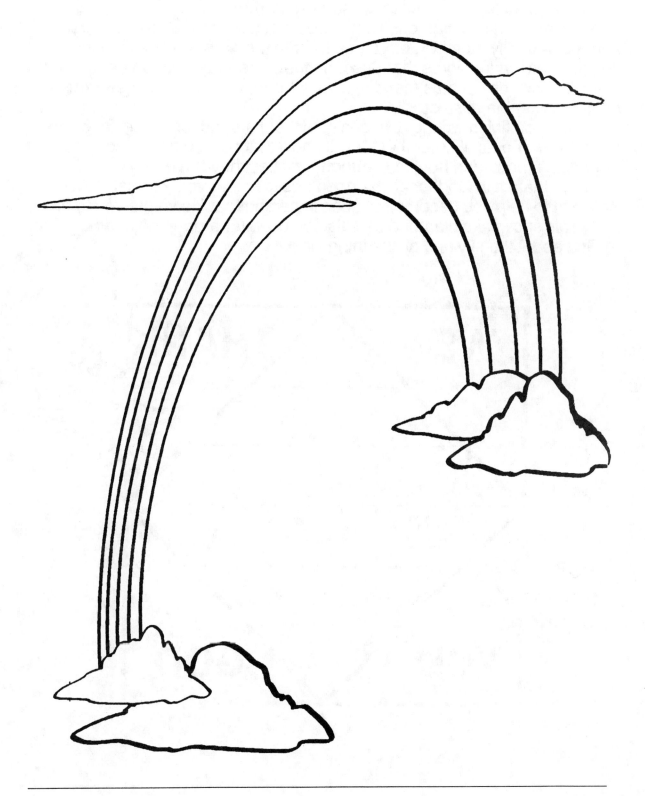

■ REPRODUCIBLE

Snowflake Diamond

In God's world everything is special, even the little snowflakes. In God's eyes, each person is special, important, and very much loved. Just as each snowflake is different, so each person is different.

You can fill in this diamond with words that tell about a special person-- a friend, a family member, or yourself. Write the person's name in the middle. Then write words that describe that person on the top and bottom lines. Use words such as friendly, loyal, strong, kind, honest, generous, cheerful, likeable, fast, etc.

When you finish writing, you can give the diamond as a special gift to your friend or a family member. It will show that person how much you appreciate them, and how many good qualities they have.

Give thanks to God for all the good qualities that you yourself have. Ask Him to help you become the special person that He wants you to be.

This snowflake diamond can also be used for a prayer of thanksgiving to God about the sky and all the things in it.

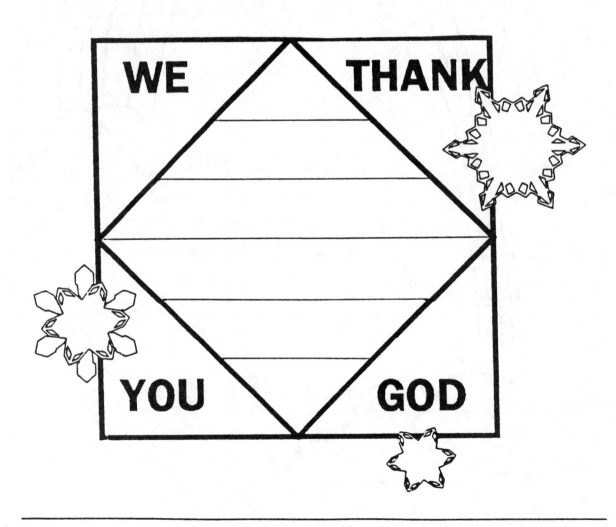

Sky Shapes

Fill in these sky shapes with your own poems, prayers, and word pictures.

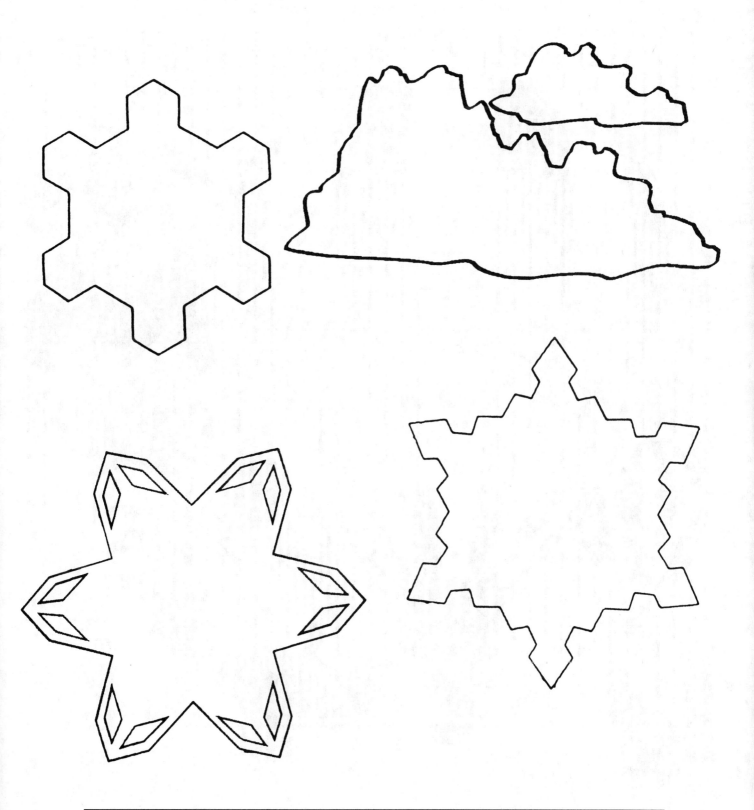

■ REPRODUCIBLE

Balloon Picture

Color in this balloon and add your own drawings and writings on the edges of the page.

HEAT INDEX CHART

TEMPERATURE

		70°	75°	80°	85°	90°	95°	100°	105°
H U M I D I T Y	0%	64	69	73	78	83	87	91	95
	10%	65	70	75	80	85	90	95	100
	20%	66	72	77	82	87	93	99	105
	30%	67	73	78	84	90	96	104	113
	40%	68	74	79	86	93	101	110	123
	50%	69	75	81	88	96	107	120	135
	60%	70	76	82	90	100	114	132	149
	70%	70	77	85	93	106	124	144	
	80%	71	78	86	97	113	136		
	90%	71	79	88	102	122			
	100%	72	80	91	108				

Do you sometimes wonder why you feel so much hotter than what the thermometer reads? The combination of heat and humidity can make us feel that way.

Forecasters have devised a heat index that tells us how hot we feel. This index tells us what the temperature feels like when heat and excessive humidity in the air won't allow moisture to evaporate.

WINDCHILL INDEX CHART

DEGREES FAHRENHEIT

WIND SPEED	45°	40°	35°	30°	25°	20°	15°	10°	5°	0°	-5°	-10°	-15°	-20°	-25°	-30°	-35°	-40°
4 mph	45	40	35	30	25	20	15	10	5	0	-5	-10	-15	-20	-25	-30	-35	-40
5 mph	43	34	32	27	22	16	11	6	0	-5	-10	-15	-21	-26	-31	-36	-42	-47
10 mph	34	28	22	16	10	3	-3	-9	-15	-22	-27	-34	-40	-46	-52	-58	-64	-71
15 mph	26	23	16	9	2	-5	-11	-18	-25	-31	-38	-45	-51	-58	-65	-72	-78	-85
20 mph	26	19	12	4	-3	-10	-17	-24	-31	-39	-46	-53	-60	-67	-74	-81	-88	-95
25 mph	23	16	8	1	-7	-15	-22	-29	-36	-44	-51	-59	-66	-74	-81	-88	-96	-103
30 mph	21	13	6	-2	-10	-18	-25	-33	-41	-49	-56	-64	-71	-79	-86	-93	-101	-109
35 mph	20	12	4	-4	-12	-20	-27	-35	-43	-52	-58	-67	-74	-82	-89	-97	-105	-113
40 mph	19	11	3	-5	-13	-21	-29	-37	-45	-53	-60	-69	-76	-84	-92	-100	-107	-115
45 mph	18	10	2	-6	-14	-22	-30	-38	-45	-54	-62	-70	-78	-85	-93	-102	-109	-117

Does your body feel chilly on a windy day? Wind has a cooling effect on the air temperature around you, causing your body to lose heat. Forecasters call this the windchill factor and have created an index that estimates the temperature you feel based on the wind speed and actual temperature.

Notes

BRIGHT IDEAS PRESS ORDER FORM
www.BrightIdeasPress.com

Returns Policy

Satisfaction guaranteed. If an item does not meet your needs, a refund minus postage will be given when the item is returned in re-saleable condition within 30 days.

www.brightideaspress.com

info@brightideaspress.com

Mail your check or money order to:

Bright Ideas Press
P.O. Box 333
Cheswold, DE 19936

Toll Free: 877.492.8081

VISA & MasterCard orders are accepted. See below for information.

Prices Good Through Until March 31, 2002
SHIPPING TABLE

Up to $50……..…….…..$6.00
$51 - $150……………….10%
over $150…………..……free

Most orders are shipped within 3 business days. Please allow 3-4 weeks from May-July.

Item #	Description	Qty.	Price	Amount
GC-100	The Ultimate Geography & Timeline Guide		34.95	
BIP-1	Hands-On Geography		14.95	
BIP-2	The Scientist's Apprentice		26.95	
BIP-3	Student History Notebook of America		12.95	
BIP-4	Over Our Heads In Wonder		9.95	
GG-102	Gifted Children At Home		24.95	
SCIAP-1	The Scientist's Apprentice Package w/ 9 reading plan books		75.00	
DW-YR1	Diana Waring Year 1 Package–Ancient Civilization		60.00	

SHIP TO ADDRESS: Please PRINT clearly

Name:

Address:

Phone: (   )

Email:

Special! FREE SHIPPING *on *orders* over $150.00*

Sub Total	
← Shipping Cost: See Shipping Table	
Total Amount Due	$

Credit Card Information

VISA/MasterCard Number   Exp. Date

Signature